Discovering Monaro

T0292207

Discovering Monaro

A Study of
Man's Impact on his Environment

W. K. HANCOCK

CAMBRIDGE
AT THE UNIVERSITY PRESS
1972

CAMBRIDGE UNIVERSITY PRESS
Cambridge, New York, Melbourne, Madrid, Cape Town, Singapore, São Paulo, Delhi

Cambridge University Press
The Edinburgh Building, Cambridge CB2 8RU, UK

Published in the United States of America by Cambridge University Press, New York

www.cambridge.org
Information on this title: www.cambridge.org/9780521104937

First published 1972
This digitally printed version 2009

A catalogue record for this publication is available from the British Library

Library of Congress Catalogue Card Number: 78-178280

ISBN 978-0-521-08439-0 hardback
ISBN 978-0-521-10493-7 paperback

Contents

Plates

Maps

Figures

Abbreviations

Agric. Gaz. (NSW)	*Agricultural Gazette* (NSW)
Agric. Gaz.	*Agricultural Gazette*
Arch. in Oceania	*Archaeology and Physical Anthropology in Oceania*
A.A.S.	Australian Academy of Science
A.D.B.	*Australian Dictionary of Biography*
A.N.U. Press	Australian National University Press
Aust. Journ. Agric. Econ.	*Australian Journal of Agricultural Economy*
Aust. Journ. Agric. Res.	*Australian Journal of Agricultural Resources*
Aust. Journ. Sci.	*Australian Journal of Science*
Aust. Journ. Zool.	*Australian Journal of Zoology*
B.A.H.	*Business Archives and History*
C.O.	Colonial Office
C.S.I.R.O.	Commonwealth Scientific and Industrial Research Organisation
Econ. Hist. Rev.	*Economic History Review*
Govt. Printer	Government Printer
H.R.A.	Historical Records of Australia
Hist. Stud.	*Historical Studies*
H. of L.	House of Lords
JRAHS	*Journal of the Royal Australian Historical Society*
JRGSA	*Journal of the Royal Geographical Society of South Australia*
Journ. Roy. Soc. NSW	*Journal of the Royal Society of New South Wales*
J. and P. Roy. Soc. NSW	*Journal and Proceedings of the Royal Society of New South Wales*
Lambie/C.S.O.	Lambie to the Colonial Secretary's Office
Leg. Co.	Legislative Council
M.U.P.	Melbourne University Press
M.L.A.	Member of the Legislative Assembly
M.L.	Mitchell Library
N.L.A.	National Library of Australia
O.U.P.	Oxford University Press
Past. Rev.	*Pastoralists' Review*
Perkins	Perkins Papers. Many volumes of excerpts from, or précis of, official documents and news items covering the first century of white settlement in Monaro (M.L., copy in N.L.A.)
Proc. Linn. Soc. NSW	*Proceedings of the Linnaean Society of New South Wales*

Abbreviations

Proc. Roy. Soc. Vic.	*Proceedings of the Royal Society of Victoria*
Quart. Journ. Agric.	*Quarterly Journal of Agriculture*
Quart. Rev. Agric. Econ.	*Quarterly Review of Agricultural Economy*
Rec. S. Aust. Mus.	Records of the South Australian Museum
S.M.A.	Snowy Mountains Authority
S.M.H.	*Sydney Morning Herald*
S.U.P.	Sydney University Press
Trans. Roy. Aust. Hist. Soc.	*Transactions of the Royal Australian Historical Society*
Trans. Roy. Soc. S.A.	*Transactions of the Royal Society of South Australia*
Vic. Hist. Mag.	*Victorian Historical Magazine*
Trans. Roy. Soc. Vic.	*Transactions of the Royal Society of Victoria*
Univ. of Chic. Press	University of Chicago Press
Vict. Nat.	*Victorian Naturalist*
V.P. Leg. Ass.	Votes and Proceedings of the Legislative Assembly

TO PEOPLE
in
MONARO

Preface

A grant by the Australian Research Grants Committee has enabled me throughout the past three years to pursue research both in Monaro and in Sydney and to buy the photostats and other material which I needed. More important still, it has given me the assistance for one year of Mr Dan Coward, whose maps, some of which are reproduced in this book, set a new standard in the cartographical elucidation of Australian historical material. I am deeply indebted to the A.R.G.C. for its support.

I am grateful to the staffs of the National Library, Canberra, and to those of the New South Wales Archives and the Mitchell Library in Sydney, for their patient and expert assistance. The keepers of records in the Lands Department and at the headquarters of the Kosciusko National Park have been most helpful to me. I am grateful to the Minister of Lands, Hon. T. L. Lewis, M.L.A., for granting me unrestricted access to these records. Miss Kyra Suthern has helped me in exploring them. At an earlier stage, Mr G. Walsh gave me valuable help.

My colleagues of the Australian National University have elected me to an Honorary Fellowship in the Research School of Social Sciences, where I occupy a room assigned to me by Professor Noel Butlin. To possess so firm a base in such friendly territory has been both a help and a joy.

In Monaro, much of my work has been done in the open air. In the summer, I have walked in the high country; in the winter, I have walked through paddocks on the tableland. Knowledgeable and agreeable companions have taught me how to use my eyes in both landscapes.

I owe more than I can say to many friends in Monaro who have explained to me how they use the land and how their forefathers used it. After we had become well acquainted, I inflicted on them an inordinately long questionnaire. Readers of this book will discover how illuminating their replies have been. Some of the replies were supported by the evidence of hitherto unexamined documents; some were supplemented by memoranda. Discussions followed. They were nonetheless rewarding when they took place on the banks of trout streams. And yet some people consider historical research a tedious occupation!

My use of the pronoun 'we' in this book is not editorial, but companionable. Not only pastoralists, but poets and scientific workers have given me generous help. So numerous have been my companions that I find it

Preface

impossible in this brief note to thank them individually. In dedicating my book to people in Monaro, I have them all in mind.

I owe my readers an explanation on a matter of technique. A bibliography, had I compiled one, would have run to inordinate length. Moreover, it would have been superfluous. My footnotes, supported by the list of abbreviations, will give the attentive reader all the information he needs about sources and their location.

W.K.H.

CANBERRA
15 October 1971

PART I

PERSPECTIVE VIEW

1. Theme

Neither word of this book's title is so simple as it seems. Let us look first at the second word.

MONARO

On 1 June 1823 Captain Mark John Currie, R.N., crossed the Limestone Plains, where Canberra now stands, and with his small party rode southwards. For two days he rode through a fine forest country intersected by stony and lofty ranges; but on the third day he saw ahead of him open, undulating, 'downy' country. There followed on 4 June a memorable encounter.

> Passed through a chain of clear downs to some very extensive ones, where we met a tribe of natives, who fled at our approach, never (as we learned afterwards) having seen Europeans before: however, we soon, by tokens of kindness, offering them biscuits etc. together with them assistance of a domesticated native of our party, induced them to come nearer and nearer, till by degrees we ultimately became good friends; but on no account would they touch or approach our horses, of which they were from the first much more frightened than of ourselves. From these natives we learned that the clear country before us was called Monaroo, which they described as very extensive: this country we named Brisbane Downs after (and subsequently by permission of) his Excellency the Governor.[1]

Mercifully, that new name did not stick. The white settlers, as they moved in, called the country Monaroo, Monera, Maneiro, Meneiro, Meneru, Miniera, Monera, and – in the fulness of time – Monaro.[2]

The name is enchanting, but the entity is elusive. We cannot, for example, feel completely sure that the white men correctly understood the aborigines who told them, so they thought, that Monaroo signified a woman's breasts, gently rounded, like the undulating downs that so

[1] Currie's account of his journey, together with map 1 on p. 4, was included in Barron Field's compilation, *Geographical Memoirs of New South Wales* (London, 1825). On 6 June Currie crossed the Bredbo river (which he mistook for the Murrumbidgee) and came within sight of the Snowy Mountains. On 7 June he began the return journey.
[2] The spellings listed are less than half the number cited by Mr W. G. R. Gilfillan in his delightful 'Discourse on the Origin of the Name Monaro', in the *Cooma-Monaro Express*, 13 December 1968.

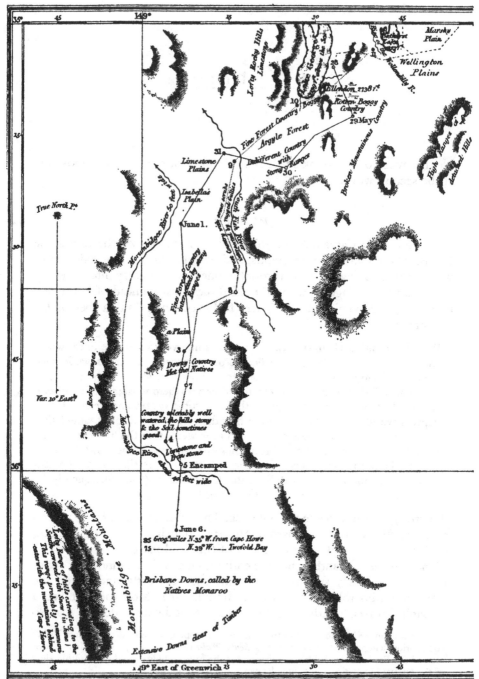

Published as the act directs 23.ᵈ April 1825, by John Mur

Map 1

4

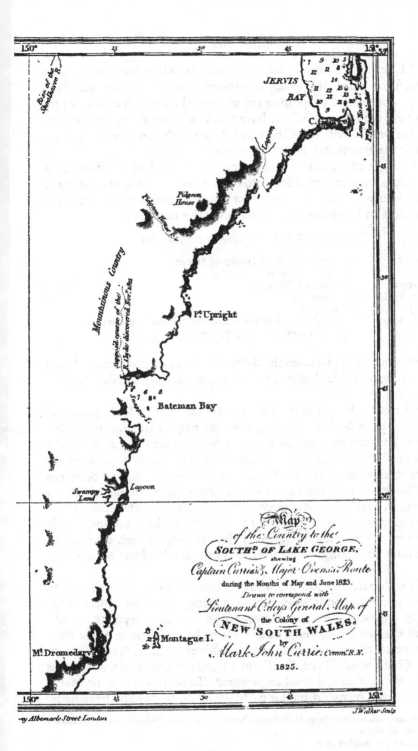

Map
of the Country to the
SOUTH.D OF LAKE GEORGE.
shewing
Captain Currie's & Major Ovens's Route
during the Months of May and June 1823.
Drawn to correspond with
Lieutenant Oxley's General Map of
the Colony of
NEW SOUTH WALES
by
Mark John Currie: Comm.r R.N.
1825.

JERVIS
BAY

Pidgeon
House

Mountainous Country

P.t Upright

Bateman Bay

Lagoon

Swampy
Land

Montague I.

M.t Dromedary

-ry Albemarle Street London

J.Walker Sculp

5

delighted Currie. Nor can we feel sure that the white men, when they pronounced the word *Mon-air-uh*, as old-timers of the district still pronounce it, were faithfully repeating the aboriginal sounds. Yet what a pity it would be if the salesmen of our busy times, who are doing all they can to make the second syllable of the word rhyme with *bar*, were to destroy at long last the traditional, beautiful pronunciation.[1]

Monaro has long since lost such chance as it ever had of achieving a precise administrative definition. Politicians and administrators have fiddled so frequently with the map that nobody quite knows where the district begins and ends, or what it contains. For example:

Fluctuating definitions of the statistical area

Up to 1873	Squattage or Pastoral District of Monaro
1874–79	Two police districts
1880–91/2	One electoral district
1892/3–1921/2	Three counties
1922/3	Three shires
1926/7–39/40	Twelve (later ten) police patrol districts, used for certain purposes in conjection with the shires

This brief summary understates the fluctuations and contradictions. Some of them are manageable; but many of them are monstrous. Monaro, as a statistical area, is a mess.

Still, we must tidy up the mess as best we can. Let us look quickly at successive maps of Monaro, starting with the map of the Squattage District (called later the Pastoral District) as it was in 1840 (map 2, p. 7). The map shows a Monaro far more widely flung than Currie's 'downland': eastwards it extends to the Pacific Ocean, southwards to the Southern Ocean. Thither, in October 1840, Commissioner John Lambie of Cooma went riding in pursuit of some cattle-rustling aborigines. He recovered no cattle and captured no aborigines; but he must surely have felt some pride of discovery, when, after riding many days through rough country, he found at long last aboriginal foot prints on the sands of the Ninety Mile Beach.[2]

Lambie's Monaro was too sprawling; it had to be cut down to size. In the mid 1840s, New South Wales itself was being cut for the benefit of the Port Philip District, which was due very soon to achieve separate colonial status and the name Victoria. The cut had to be made along a line from Cape Howe to the nearest source of the River Murray. The Surveyor General of New South Wales, Sir Thomas Mitchell, entrusted that arduous survey to his capable and resolute assistant, T. S. Townsend. In March and April 1846, Townsend and his party, sweating by day and freezing by

---

[1] The *bar* pronunciation of the second syllable has been adopted by General Motors Holdens for the car they call Mon*a*ro.

[2] Lambie/C.S.O.: 9 October 1840.

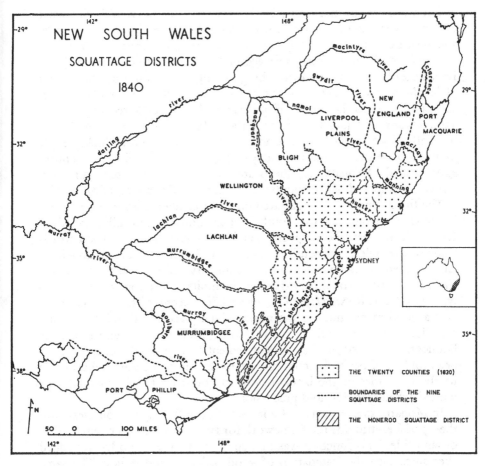

Map 2

night, forced their way through the tangled ranges and the dense scrub of the borderland. Townsend's line was not the line that Mitchell wanted; but higher authority had ordered it. On the map it looks razor sharp.[1]

No deliberate plan was ever made to amputate the Squattage District on its Pacific flank. On the contrary, in the mid-1830s, Governor Bourke envisaged an economic union between the eastern ports and the western pastures; in the mid-1840s, Benjamin Boyd did all he could to make that

---

[1] Townsend to Mitchell, 14 and 16 April 1846 and 1 December 1850; Mitchell to Townsend 30 November 1850. Mitchell, if his instruction had allowed him, would have had the line drawn along 'the most clearly defined boundaries', from Cape Howe to the Murrumbidgee, at or near Wagga Wagga. This line would have given to the Port Philip District a large increment of territory.

vision a business proposition.[1] To explain how, and why, and with what consequences, the opposite happened, would require rigorous economic and historical analysis. If any official person of the 1830s or '40s ever attempted such an analysis, no trace of the attempt has survived. John Lambie took it for granted that the gazetting of a county on the coastal plain diminished to the same extent the area of his own territorial responsibility. Thus the dream of a Greater Monaro flickered out. To be sure, it was never quite forgotten; at the end of the Second World War, a few enthusiasts were trying to kindle it again. They had little luck. People in Monaro see their homeland today as a guidebook decribed it a century ago: 'a land of squattages, sheep and wool, "downs", mountains and rivers...composed of the counties of Wellesley, Wallace and Beresford'.[2]

The three counties were gazetted on 21 December 1848 (map 3, p. 9). Their length from north to south is approximately 100 miles and their breadth from west to east approximately 50 miles; their combined area is 5,400 square miles. The southern and eastern boundaries of this area have already been indicated. The western boundary follows the alpine watershed. The northern boundary is at the old 'limits of location';[3] consequently, a traveller on his way from Canberra to Cooma enters the three-counties area a good many miles to the north of the place where the first white traveller heard for the first time the word 'Monaroo'. By and large, the boundaries gazetted in 1848 are convenient. The northern one, we have been told, is of particular convenience to the birds; on the Canberra side lie the territories of the black-backed magpies; on the Cooma side the white-backed magpies hold possession.[4]

In Monaro, as everywhere else in New South Wales, the counties, with their parishes, provide the framework for land survey and the registration of land title. This function was first spelt out in 1825[5] and has remained ever since then the foundation of good order in land policy. The same foundation might conveniently have supported the fabric of local government; but this did not happen. Early in the twentieth century, the country districts of New South Wales received a belated first instalment of self-government. Monaro was divided into three shires. The external boundary

[1] H.R.A. I, vol. XVII, pp. 468ff., Bourke to Stanley, 4 July 1834.
[2] Lambie/C.S.O.: 24 March 1845; Alfred MacFarland, *Illawara and Monaro Districts of New South Wales* (Sydney, 1872), pp. 67–8.
[3] See below, pp. 41, 82.
[4] The late Mr Ernie Quodling, as cited in the *Cooma–Monaro Express* of 19 June 1959, pin-pointed Michelago as the place where a traveller on the main road can see the white-backed magpies take over from the black-backed magpies. The observations of the present writer, for what they are worth, support the Quodling proposition.
[5] H.R.A. I, vol. VI, pp. 434ff., Bathurst to Brisbane 1 January 1825. This most important despatch gave orders for survey and the division of the entire territory of New South Wales and Van Diemen's Land into counties, hundreds and parishes. We still have the counties and parishes, but have got along quite well without the hundreds.

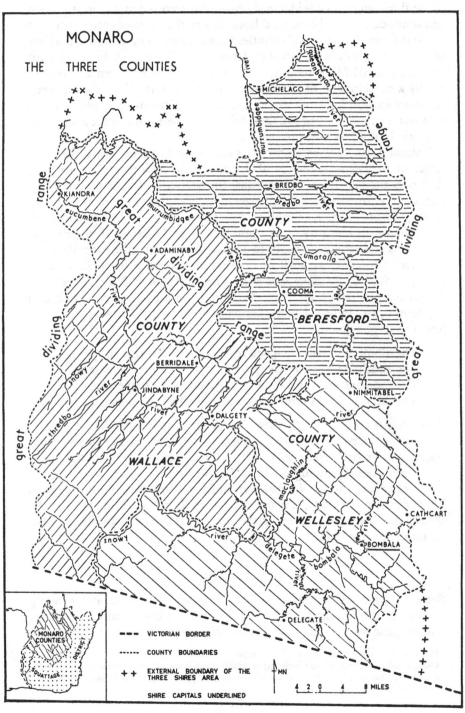

Map 3

9

of the three-shires area is identical, except for some squiggles in the north and south-east, with the external boundary of the three-counties area; but inside the area the shires and counties have different boundaries. The shires are named Bibbenluke, Snowy River and Monaro – a little Monaro within Monaro! More recently, the larger Monaro has been restored to life by a combination of the three shires: for certain limited purposes,[1] this combination has been named the County Council of Monaro. If the administrators were to put all their different Monaros on to the same map, it would look like a crazy pavement.

Natural scientists, by contrast, make unambiguous maps. Using a larger or smaller scale as suits their purpose in each particular instance, they map the geological foundations, the surface soils, the altitudes, the rainfall, the temperature, the vegetation, the distribution of Bogong moths, the distribution of rabbits. Ecologists, in their attempt to understand the whole interplay of a natural habitat, make use of all these maps. Historians and ecologists ought to keep in close touch with each other. As our study proceeds, we shall make frequent reference to A. B. Costin's book, *A Study of the Ecosystems of the Monaro Region of New South Wales.*[2] Costin accepts, with one or two variations, the three-counties definition of Monaro, and in broad terms defines as follows its natural subdivisions (see map 4, p. 11):

1. The alpine tract: above 6,000 feet (=above the tree line).
2. The subalpine tract: from 5,000 to 6,000 feet, on average.
3. The montane tract: from 3,000 to 5,000 feet, on average.
4. The tableland tract: from 2,000 to 3,000 feet, on average.

An ecologist certainly needs these four tracts, but an historian may make do with two: alps and tableland. This book will tell the tale of two landscapes. Or – more precisely – it will tell the tale of the men who have made, and are still making, these landscapes.

We need explore no further the shifting definitions of our region. Monaro, let us conclude, is marcher country. It has no fixed frontiers. Its three counties are home base; but the historian is free to roam beyond base, as did the people in his book.

DISCOVERING

Some gifted draftsmen of the First Fleet made accurate and pleasing pictures of the hills and valleys around Sydney; but a century went by, a great geographer once said, before any practitioner of his profession

---

[1] Most notably, the control of electricity supply.

[2] Govt. Printer, Sydney, 1954. Costin departs from the three-counties definition in the north-east, where his map (*Ecosystems of the Monaro Region*, pp. 3 and 21–3) shows an extension towards Queanbeyan.

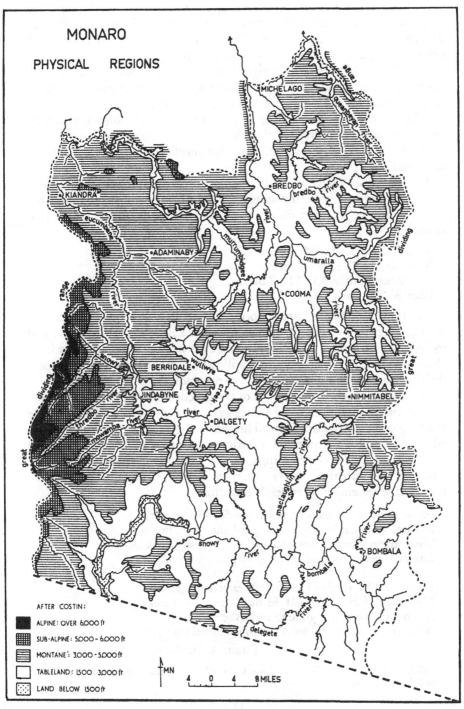

MONARO

PHYSICAL REGIONS

AFTER COSTIN:

■ ALPINE: OVER 6,000 ft

▤ SUB-ALPINE: 5,000 - 6,000 ft

▥ MONTANE: 3,000 - 5,000 ft

□ TABLELAND: 1,500 - 3,000 ft

⬚ LAND BELOW 1,500 ft

↑ MN

4 0 4 8 MILES

Map 4

really knew those hills and valleys.[1] A soil chemist or a botanist, a painter or a poet, might equally well have said that; they would all agree that the discovery of Australia, or of any Australian region, is not a once-for-all achievement, but rather is a continuing effort, whose end – if ever there is an end – still lies far beyond sight. Further, they would agree that this effort has its own logic and rhythm; starting with the scrutiny, collection and classification of objects on the surface of the land, it proceeds by searching the deeper levels. Let us give a name to the discoverers who see their task thus; whether they be men of science, speaking to our intellects, or men of art, speaking to our affections, let us call them 'the observers'.

There are discoverers of another stamp whom we shall call 'the practical men'. A tombstone at Campbelltown commemorates in rustic English a practical man of the First Fleet, the convict James Ruse, Australia's first farmer.

> My Mother Reread Me Tenderly
> With Me She Took Much Paines
> And When I Arrived In This Coleney
> I Sowed The Forst Grain.

Practical men in Monaro grow wool and meat in preference to grain; but their task has been essentially the same as James Ruse's task; they have had to learn the hard way, by trial and error. Nowadays, to be sure, they have abundant scientific advice, but they still have to make their own decisions, their own mistakes and their own 'improvements'.

Improvement is the watchword of the practical men; but the observers do not always accept the word at face value. A century and a half ago, the Polish geologist Strzelecki forced his way along the rough south-westerly edges of Lambie's Squattage District and saw the land aflame. The squatters, he told Governor Gipps, were not improvers, but spoilers of the land.[2] In our own time, many observers in many Australian neighbourhoods have reiterated Strzelecki's indictment.

> For the Earth, our mother, at last has found a master;
> She was slow and kindly, she laughed and lay in the sun –
> Time strapped to his wrist, he made the old girl work faster,
> Stripping her naked and shouting to make her run.
>
> .   .   .   .   .   .
>
> He cracked his stock-whip: that characteristic gesture
> Made dust of the plains and the hurricane bore it away
> A thousand years had gone to make the pasture
> Which the wind or the flood destroyed in a single day.
>
> (From A. D. Hope, *Toast for a Golden Age*)

[1] Griffith Taylor, *Sydneyside Scenery and how it Came About* (Angus and Robertson, 1958), p. 2.
[2] See below, pp. 57–9.

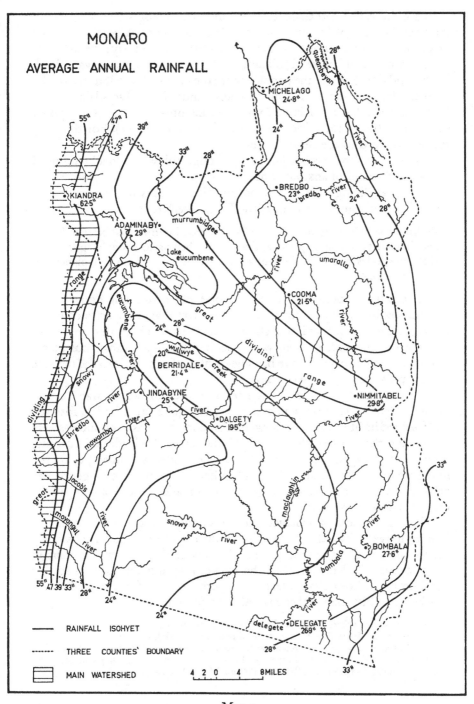

Map 5

In a book called *The Great Extermination*, six biologists have spelt out the poet's statement in detail and, at times, in anger.[1] A painter, Russell Drysdale, moves us, not to anger, but to pity and anxiety; we ask ourselves, as we stand before his pictures of wasted humanity in a waste land, 'Is the land doing this to the people, or are the people doing it to the land?'

Australia, of course, has land of many kinds. From Drysdale's desert to Monaro is a far cry. Alpine Monaro holds pre-eminence as a supplier of water to the world's driest continent (see map 5, p. 13). For this very reason, we need to use Monaro well.

Misuse of the land has become for all thoughtful people in all regions of the world a matter of anxious concern. It is currently on the agenda of the International Biological Programme. Introducing this programme to his Australian colleagues, Sir Macfarlane Burnet declared:

> Perhaps I should state my own conviction that international science has or should have two basic postulates: that all men belong to a single species and that the species will have the whole of its future history on this planet and nowhere else.[2]

Some scientists would dispute the second postulate; but none would dispute the first. Burnet went on to say in effect that the species could all too easily turn the planet into a stinking space ship. Those thin envelopes of air, water and soil, without which neither men nor animals nor plants nor insects could survive, have become painfully vulnerable to the massive powers possessed by modern man.

We have been warned. Still, we need not lose our nerve. Man has too often spoiled the land on which he lives; but quite often he has fostered and fortified its life-sustaining powers. Sometimes he has been successively the spoiler, the restorer, the improver. The terraced hillsides of Tuscany, Italy's best managed agricultural province and loveliest landscape, are the living embodiment of this rhythm.[3] What the human intelligence has achieved in Tuscany it can achieve, not by imitation but by a comparable creative effort, in Monaro.

What the chances of achievement are we shall see more clearly as our inquiry moves forward from prehistory through history to the anxiously debated issues of present time. The inquiry is this: 'How has man in Monaro used the land on which he lives?' Looking for answers to this question could keep a large team of experts at work for a long time; in this book an historian makes his bid for a place in the team.

[1] A. J. Marshall and others, *The Great Extermination* (Heinemann, 1956).
[2] See *Biology in the Modern World* (symposium of the A.A.S., 1967), pp. 1–3.
[3] Nearly half a century ago, the present writer made acquaintance with the makers, menders and minders of the Tuscan landscape. See his *Ricasoli* (Faber, 1926), ch. 3 and his paper 'Italian Métayage', reprinted in *Politics in Pitcairn* (Macmillan, 1947).

2. The first discoverers

Casually, Captain Mark Currie recorded the name Monaroo. He proposed in its place an import from Scotland, but nonetheless we feel grateful to him for having asked those timid Aborigines what they called their country. We should feel more grateful to him still if he had gone on to ask them what their tribe called itself, how large the tribe was in numbers and extent of territory, and how the men and women of the tribe earned their living on the land by use of their spears and nets, their digging sticks and fire sticks. Today, half a century after the death in Cooma of Biggenhook, the tribe's last survivor, questions such as these seem to us important. Captain Cook would have thought them important; but Captain Currie was an incurious man.

His main concern, we may suppose, was to find new pastures for the colony's cattle and sheep. Apart from that, cultivated persons of his time seldom felt any strong urge to ask questions about Aboriginal life, because the answers were already manifest – so they thought – to any intelligent European. If their frame of thought was theological, they took it for granted that God, having created all human creatures as a single family, desired the conversion of Aboriginal peoples to Christian doctrine and civilised living: alternatively, they took it for granted that God, having chosen some men for salvation and others for damnation, had damned the Aborigines. If, on the other hand, their frame of thought was secular, they envisaged an original state of nature within which – some of them maintained – noble savages lived an uncorrupted life; but others maintained that nasty savages in their state of nature lived a brutish life. The approach of Europeans to Aboriginal societies, whatever its initial starting point, was almost invariably slanted towards one or the other of these conclusions. In New South Wales, the optimistic approach was fashionable to begin with; but the pessimistic one soon superseded it. Both approaches were an impediment to scientific observation and human understanding.[1]

Moreover, when New South Wales was founded, the chronological system of Moses, as expounded by Archbishop Ussher, still remained for many people the frame of reference. Here was a main impediment to

[1] See D. J. Mulvaney, 'The Australian Aboriginees, 1606–1929: Opinion and Fieldwork' in *Historical Studies, Selected Articles* (M.U.P., 1964). See also Bernard Smith, *European Vision and the South Pacific, 1768–1850* (O.U.P., 1960).

intelligent study of Aboriginal society: given that Europeans and Aborigines shared a common descent from Adam, no rational explanation could be discovered of how, within the short span of 6,000 years, they had become so conspicuously dissimilar. Unless and until the Mosaic chronology was demolished, the study of Aboriginal life could get no further than superficial fact-collecting.

During the middle decades of the nineteenth century, geological and biological research steadily eroded the Mosaic chronology. In 1859, Charles Darwin's book, *The Origin of Species,* gave it the *coup de grâce.* From that time onwards, empirical observers of savage society – 'savage' was then a fashionable word – had the backing of systematic theory. In Australia as elsewhere, the field-workers asked purposeful and pointed questions, not only about the sizes and shapes of Aboriginal heads, but also about the beliefs inside those heads; not only about the material fabric of Aboriginal life, but also about its social fabric. From this systematic questioning, large increments of knowledge accrued.

Nevertheless, the increments might have been larger still had not the questions been so sharply slanted. The philosophy of progress pervaded, and at times distorted, sociological inquiry. Seen in the light of this philosophy, human history became the successive stages of a forward movement from simplicity to complexity, from savagery to civilization. In the 1830s Auguste Comte had identified three stages – the theological, the metaphysical, the scientific. In the 1890s Sir James Fraser similarly identified three stages – the magical, the religious, the scientific. In classifications such as these, the Aborigines found their niche as 'a survival' of Stage One. Survivals and stages, no doubt, have their uses; but for the historian they are not enough. He wants to meet the Biggenhooks.

The social anthropologist feels a similar want. To satisfy it, he goes to live inside the society which he has chosen as his object of study. 'Participant observation' is his slogan. It is a good slogan, except for one defect: the accent falls too heavily on the society as it functions here and now. But there may not be any 'here and now'. The Biggenhooks of Monaro are all dead. A reflected image of the life they once lived may still survive in the written or printed observations of white Australians; but if we seek more knowledge of their past, we have to dig for it. A century ago, the digging of an enthusiastic German brought to the light of day the Bronze Age past of Troy and Mycenae; in this century, archaeologists have been at work in many countries digging up the past of Stone Age man. In 1929, two Australian diggers, Herbert Hale and Norman B. Tindale, gave a sensational hoist to our knowledge of the Aboriginal past. At Devon Downs on the Murray River they established the fact of human occupation for upwards of 5,000 years. Moreover, by digging up material evidences of a succession of cultures, they demolished for all time the myth of an inert Stage One. Black Australians could no longer be en-visaged in the lump as an undifferentiated, unchanging people, totally isolated from any infusion of new blood and new experience, unconcerned and unable to strike out for themselves along new paths of need and opportunity.[1]

Notwithstanding some controversies that have ensued since 1929 – should we, for example, envisage five successive cultures, or two, or three? – the excavation at Devon Downs remains a majestic landmark of dis-covery. A second landmark is the publication in 1952 by an American physicist, W. F. Libby, of a method – the radiocarbon method – of dating archaeological deposits with quite remarkable precision.[2] A third land-mark is the inauguration in 1961 of the Australian Institute of Aboriginal Studies. Since then, our knowledge of the Aborigines has surged ahead.[3] For example, we can now feel sure that they were already in Australia 30,000 years ago, and we can look forward to the demonstration very soon of still earlier dates.

Unfortunately, we do not yet have any dates for Monaro. Archaeolo-gists have barely started to dig there. Still, they are digging on the Pacific Coast close by and have already given a date – 20,000 years ago – to a site at Burrill Lake.[4] May we not look forward to the discovery in Mon-aro of sites comparably old? This expectation would be reasonable, if we could feel sure that the climate of Monaro was warm enough then for the Bogong moths, and for the men who feasted on the moths. But, in the study of past climates, the zones of uncertainty still remain wide. We believe that the Aborigines came to Australia from Asia during the fourth and last ice age of the Pleistocene; that about 10,000 years ago – when the land bridge to Tasmania was drowned – the ice caps of Antarctica were dwindling rapidly; that the Alpine summits of Monaro must by then have become hospitable to moths and men. What we do not yet understand are the pauses and oscillations in the general trend from a colder to a warmer climate. Even so, it seems to be widely agreed nowadays that the area of glaciation on the Snowy Mountains was never larger than about 50 square kilometres. As soon as the ice began to melt, moths and men may well have become summer visitors to the high country; below the subalpine contours, marsupials and men may well have been roaming, even during the ice age. Some day, let us hope, these doubtful issues will be decided;

[1] H. M. Hale and N. B. Tindale, *Notes on Some Human Remains in the Lower Murray Valley*, Rec. S. Aust. Mus., 4 (1930): 145–218.

[2] Twenty years ago, datings of more than 20,000 years were not attempted; today, the period has been more than doubled, although the margin of error increases with distance in time.

[3] See D. J. Mulvaney, *The Prehistory of Australia* (Thames and Hudson, 1969) and D. J. Mulvaney and J. Golson (eds.), *Aboriginal Man and Environment in Australia* (A.N.U. Press, 1971).

[4] See R. J. Lampert's contribution in Mulvaney and Golson, *Aboriginal Man and Environment in Australia*, pp. 114–32.

in the meantime, let us by a bold guess split the difference in time between the sinking of the Tasmanian land bridge and the radiocarbon date at Burrill Lake. That rough arithmetic would bring man into Monaro about 15,000 years ago. By reckoning each generation of men as 25 years, we could say that 600 generations of black Australians had lived in Monaro before the white Australians took possession there.[1] So far, only six generations of white Australians have lived in Monaro.

The tribal territories of the black Australians are shown on map 6, p. 19.[2] The unshaded area below the 1,500 feet contours has been included, because we have found within it the springboard for our chronological plunge; but the tribes there will have no place in our story; by and large, they stayed put on the coastal plain.[3] The shaded area indicates our field of study. It includes the territories of all the tribes, both north and south of the colonial border, which had ready access to the alpine summits to which the Aborigines repaired in large numbers during the summer months, for a reason which will later become clear.

So far as we can, we must paint the picture of these black Australians at work on the land to earn their living. The attempt may seem pretty hopeless, seeing that it depends in large measure upon the recorded observations of white Australians whose main concern was to earn a living for themselves. At first, the whites took notice of the blacks chiefly as a menace or a hindrance to themselves; a little later, they looked for ways and means of making the blacks useful; still a little later, they dismissed the blacks altogether from their calculations. There were, of course, exceptional white men, like Murray of Yarralumla and his young friend Stewart Mowle; Murray would have wished the blacks still to live their old independent lives, but did not see how this could be; Mowle liked listening to them singing, but was able to remember and record only a few words of their songs.[4] Later in the century, the thirst for ethnographical and linguistic knowledge grew sharper; but by then it was almost too late. Our picture of black Australians earning their living on the land in

[1] More than 600, if we follow J. B. Birdsell's estimate of 18 years for an Aboriginal generation.
[2] See *Distribution of Australian Aboriginal Tribes: A Field Survey* by Norman H. Tindale, reprinted from *Trans. Roy. Soc. S.A.* LXIV (1) (1940), 140–231. Accompanying this most important paper is a map showing the distribution of all the Australian tribes. The south-eastern corner of the map is reproduced on p. 19, with revisions which Dr Tindale has kindly made available to the author.
On the use of the terms 'white' and 'black', here and later, see below, p. 27, note.
[3] See below, p. 69. The coastal tribes had two good reasons for staying put: first, they had plentiful supplies of good food close at hand; secondly they were on terms of hostility with the tribes inland. See, e.g., 'George Augustus Robinson's Journey into South Eastern Australia 1844', edited by George Mackaness in *Trans. Aust. Hist. Soc.* XXVI (1942), 332–4. However, the inter-tribal hostility which Robinson noted may possibly have been the consequence of post-1788 pressures. A. W. Howitt and other observers of the later nineteenth century recorded instances of friendly relations between coastal and inland Aborigines.
[4] S. M. Mowle, *List of Aboriginal Names in the Southern Highlands* (Sydney, Govt. Printer, 1891).

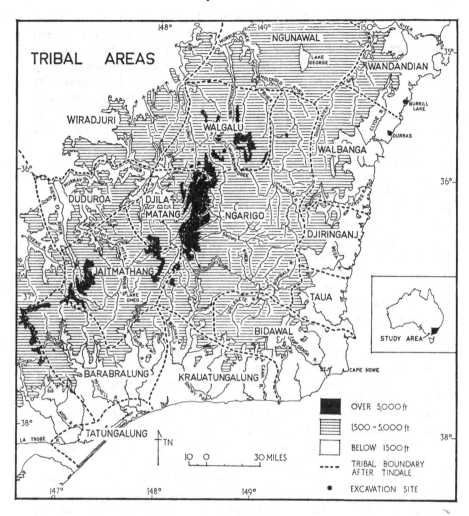

Map 6

Monaro would be empty indeed were it not for the advent, almost at the last hour, of two remarkable men, A. W. Howitt and Richard Helms.

These two men are a reminder to us that the classifications proposed in the previous chapter are for initial convenience only; in real life we continually meet people whose faculties – artistic and scientific, practical and observational – are variously intermingled. Both Howitt and Helms were resourceful practical men; both were keen and critical observers. Howitt possessed a good educational grounding – although no university degree – both in languages and the natural sciences; he read and spoke German; he had a trained eye for rocks and soils and plants; he was a competent draftsman. After voyaging to Victoria from England during the gold rush he lived for a time rather precariously; but he taught himself bushcraft and before long achieved fame as an explorer. Thereafter he became the police magistrate of Gippsland and from that base achieved more enduring fame as an ethnologist. Helms, like Howitt, possessed no university degree; but he possessed all the qualities of that glorious race, now sadly diminished, of field naturalists – the love of learning, the habit of precise and thoughtful observation, the zest for exploration, the explorer's practical capacity. In 1858 he took ship from Hamburg, landed in Melbourne and went into business with his cousin, selling cigars. Three years later he went to the South Island of New Zealand, where he earned his living by mending watches, pulling out teeth and other useful employments. Whenever he could break free from the tasks of earning a living, he made rough and solitary expeditions to the zoologically unexplored west coast; numerous shells and a dozen or more insect species that he collected there are named after him. In 1888 he crossed the Tasman Sea again to take up a position in the Australian Museum; the following year he made his first expedition to the Kosciusko country. Before his death in 1914, he had followed half a dozen other diverse employments attractive to him as a field naturalist; but we need follow his career no further than the Australian Alps. The fruits of his explorations there were four illuminating papers – on the evidences of glaciation at Mt Kosciusko; on the topography and fauna of the Snowy Mountains; on the snow leases of the white Australians; on the way of life formerly led by the vanished tribes of Aborigines.[1]

In his *Anthropological Notes,* Helms uses two sources of information: first, the evidence submitted to him by old settlers who had known the Aborigines before they became – to quote Captain Currie again – 'domesticated': secondly, the evidence of his own eyes, as exemplified by his descriptions and drawings of the implements that he had collected in river shingle below Jindabyne, and of the grave that he had opened in the Mowamba valley. Sociological theory and controversy, if he had any awareness of it, held no interest for him; in his own cool style as a field naturalist he reported such facts as he had gathered about intertribal relationships, tribal government, the rules of marriage, the rearing of children, the coming of age and the tribal initiations of boys and girls, sickness, old age, death, mourning and burial. These patterns of social custom would not here concern us, were it not that they intermingle with patterns of the natural environment. In Helms' picture of the Aborigines there is intermingling everywhere – of the living with the dead, of the men with other living creatures, of the men and the creatures with spiritual

[1] On Helms, see Charles Hadley in *J. and P. Roy. Soc. NSW* XLIV (1915), 11–14. On Howitt, see Mary H. Walker, *Come Wind Come Weather* (M.U.P., 1971).

beings. All these relations are personal. The black people of Monaro, we feel, are living with the land, not merely living on it.

Nonetheless, they had to be hard-headed, if not ruthless, in their struggle to earn a living. Helms gives us a detailed account of their material equipment and of the uses to which they put it. He tells us how they made fire by rubbing together two pieces of the seed stalk of the grass tree; how they made camp ovens; how they made belts, cloaks and mats from the skins of possums and kangaroos; how they stripped bark from the trees and used it to make their simple shelters and not-so-simple canoes; how they stripped, separated, span and plaited kurrajong fibres to make their finely meshed nets for catching insects; how, for the pursuit of larger game and for battle, they made sharp-edged axes and other stone implements. He gives us a list of their weapons and implements, whether made of stone, wood, reed or bark: axes; boomerangs of two kinds; spears of three or four kinds; woomeras for throwing the lighter spears; clubs, digging sticks that could at need be used as clubs; nets and bags and water-carrying vessels. With this equipment and with a realistic division of labour between the sexes, they lived busy lives, hunting and gathering animals large or small, birds and birds' eggs, reptiles, fishes, insects, seeds and roots. Possum meat, probably, was their main culinary standby, except during the summer months, when they grew fat and glossy – and lazy as well, some white people said,[1] by feasting on the Bogong moths.

Helms's account of the ways of moths and men was first-rate, but when he wrote it the men were no longer in the mountains; the crows by then were the only hunters and could be seen, when anything frightened them, rising in their thousands from the cracks and caves, like the seabirds that rise from a lonely, rocky island when a sudden shot disturbs them. Consequently, for further information about the men, we must track down earlier witnesses. For further information about the moths, we can call to witness scientific observers of our own time, starting with Dr I. F. B. Common, who spent three successive summers during the 1950s in the Brindabella mountains near Canberra, observing, measuring and recording the behaviour of *Agrotis infusa*.[2] On the basis of his report, a good many firm statements can be made. The spring migration of the moths to the mountains begins in late October or early November. On their flight inwards and upwards, the moths feed freely on the spring blossoms; but in their lofty summer residences they take little or no food. They do not need it, for they do not breed in the mountains or take much exercise there. They rest and conserve their strength, clustering close together in the dark

[1] E.g. Edward Bell in *Letters from Victorian Pioneers*, ed. T. F. Bride (Govt. Printer, Melbourne, 1899), p. 171.
[2] I. F. B. Common, 'A Study of the Ecology of the Adult Bogong Moth *Agrotis infusa*', in *Aus. Journ. Zool.* II, no. 22 (1954), 223–62.

crevices and caves, heads down, each moth gripping the rock surface with its fore tarsi and resting its middle and hind legs on the back of the moth behind it. As the summer draws towards its close in late February and March, the clusters dwindle as successive contingents of moths set out on the return journey, near or far, to their winter residences on the heavy soils of the slopes and plains.[1] There they mate; there the larvae grow into ravenous cutworms, feeding on the crops of infuriated farmers, until spring returns, wings are spread, and the merry game begins again.

Dr Common suggests a closely reasoned explanation – 'aestivation' and 'facultative diapause' are his key words – of these extraordinary migrations; but we have to push ahead with our own inquiry, in which the men take precedence over the moths. To the Aborigines, the inward flight each spring must have come almost literally as a gift from heaven, for there were millions of moths and each of them was an animated fat bag.[2] The fat was nourishing; moreover, it was tasty. This, at any rate, was the considered opinion of Mr Robert Vyner, who in the summer of 1865 made two successive expeditions to the mountains under the guidance of an Aboriginal whom he called Old Wellington. On one of these expeditions, he ate about a quart of moths and so much enjoyed their sweet, nutty flavour that he would have liked, he said afterwards, to have had another feed. Like Helms, who described in detail all the motions the Aborigines went through between catching their meal and eating it, Vyner gave them high praise as cooks.[3]

T. S. Townsend, by contrast, denounced them as incendiaries.[4] In March 1846, while he was making his survey of the colony's southern border, the blacks were on the mountain tops, feasting on the Bogongs. He put the blame on them for the burnt scrub that he had to drive his bullocks through and for the dense clouds of smoke that made it almost impossible for him to get his angles on distant points. We need not take every item of his statement as gospel truth; he reported fires alight on the timbered slopes *below* him, not in the high country, where the blacks were hunting and feasting. For geographers, botanists and anthropologists of our own time, this distinction is important. Most of them would acquit the Aborigines on the charge of setting the snowgums and the alpine pasture alight, but would pass a different verdict on their use of their

[1] *Ibid.* p. 266, a map showing the most important breeding grounds, so far as they are known.
[2] *Ibid.* and Helmut Reim, *Die Insektenbenahrung der Australischen Ureinwohner* (Berlin, 1962), pp. 32–3. In males, the proportion of fat to dry body weight is at times well above 70 per cent; in females, above 60 per cent.
[3] Vyner's raptures were fully reported in 1903 by W. W. Froggatt, in his paper 'Insects used as Food by the Australian Natives' in *Science of Man* (Sydney), VI(i), 12.
[4] See p. 6, above. Townsend's tirade is well known; it appeared in print on pp. 227–8 of W. B. Clarke's *Researches in the Southern Goldfields of New South Wales* (Sydney, 1860). A vivid early account of the moths and men is in George Bennet's *Travels, in New South Wales* . . . (London 1834) vol. I, pp. 265–76.

22

firesticks at lower levels. Some of them have gone so far as to argue that the grasslands and open woodlands, such as those that Captain Currie saw on the tableland, are not a product solely of the interactions of climate, soil, plants and animals, but have been shaped by palaeolithic man.

Arguments of similar intent are currently matter of debate in every country where scientists and humanists are seeking new knowledge of the millenial interplay between man and nature. Modern man, notwithstanding the technological triumphs that dazzle and at times terrify him, is not the first revolutionary; to speak of 'the neolithic revolution' has become a commonplace. But we have not, as yet, found room in our vocabulary for a palaeolithic revolution. Until recently, it has been widely assumed that men of the old stone age, in Australia as elsewhere, made niches for themselves within the natural environment, but made no measurable impact upon the environment itself. Now, however, the pendulum of research and explanation is swinging the other way. For example, it used to be taken for granted that the giant marsupials of Australia fell victims to a change of the climate; but today it is strongly argued, although not as yet proved, that the Aborigines destroyed them. Many comparable examples from many regions of the world incline us towards the belief that palaeolithic man, puny though his equipment may seem to us, produced many notable changes upon the surface of the earth. For one thing, he had tens of thousands of years in which to produce them; for another thing, to say that his equipment was puny is to beg the question. We must not so casually dismiss the parable of Prometheus.

A quarter of a million years ago, at the very least, forerunners of *Homo sapiens* were tending fire on the hearth in a cave near Pekin. We are unable to assert that they had kindled the fire; but we may feel sure that they cherished it and tried to take it with them if and when they went travelling to a far country. We may feel just as sure that the generations of men who came later in time made persistent and effective use of fire after they had discovered the art of kindling it for themselves. We reach this conclusion by analogical reasoning from the observed practices of the most primitive peoples in all the continents; they used fire – and still use it, where they survive with their traditional way of life – for their hunting drives and for thinning the forest and scrub, both to make their own movements easier and to make fresh pasture for the wild animals. Their firesticks, some people argue, have been the main makers of grassland and savannah in many regions of the world.[1]

This argument has won little support, until quite recently, among students of grassland in Australia. By and large, they have kept stone age man with his firestick out of the picture. Sometimes they have had a

---

[1] See, for example, the papers by H. C. Darby, Carl O. Sauer and Omer C. Stewart in *Man's Role in Changing the Face of the Earth*, ed. William L. Thomas (Univ. of Chic. Press, 1956).

sufficient reason for keeping him out. For example, in the Tantangara district of Monaro, cold air draining on still nights from the hills into the hollows has reversed the customary correspondence between vegetation and altitude – trees grow on the warmer summits and slopes; but in the hollows no tree grows.[1] Even so, it is a far cry from these subalpine frost hollows, with their narrow exits, to the spacious, undulating, 'downy' country which Captain Currie marked on his map: in that country, some other explanation of treelessness has to be looked for. It need not be a single explanation; most Australian botanists have enumerated various interacting natural causes – rainfall, temperature, the lie of the land, the chemical composition of rock and soil. A botanist who knew Monaro well, R. H. Cambage, laid most of the emphasis upon the last-named natural cause: the grassy plains are treeless, he said, because their soils are basaltic: if ever an isolated knob has trees on its summit, it is sure to have a plentiful supply of silica in its rock and soil.[2] A Victorian botanist, R. H. Patton, offered a similar explanation of the grasslands south of the Murray. 'It is one of the great assets of this country', he asserted, 'that at the time of white settlement, its native vegetation was wholly related to its physical environment and was undisturbed by man or animals.' Patton took particular pains to exclude the activities of man from any part at all in making the grasslands: the Aborigines, he asserted, unlike their white supplanters, had started very few fires.[3]

That assertion was too impetuous a leap from botany into history. Botanists nowadays start from the proposition that Australian vegetation is highly adapted to fire;[4] then they seek the explanation. Some single-minded expositors of 'natural' causation may still lay the main emphasis upon lightning strike; but most people now believe that the Aborigines kindled a very large number of the fires which throughout so many thousands of years made so large an impact upon plant life in Australia. This belief rests upon cumulative evidence. The sea explorers reported fires and 'smooks' almost everywhere along the Australian coast, and accepted them as evidence of human habitation. The land explorers, almost without exception, saw the Aborigines constantly and purposefully at work with their firesticks. The graziers, as time went by, saw and reported the return of trees and scrub to country which had been open when they first took possession of it. For this most unwelcome change, no

[1] See R. M. Moore, 'Natural Phenomena and Microclimate' in *Proc. Canberra Symposium Climatology and Micrometeorology* (UNESCO Paris, 1958).
[2] Presidential Address to the Linnaean Society of New South Wales, 25 March 1925, pp. xix ff.
[3] Reuben T. Patton, 'The Factors controlling the Distribution of Trees in Victoria', *Proc. Roy. Soc. Vic.* XLII (n.s.), pt. II (1930), esp. 161. Cf. *Proc. Roy. Soc. Vic.* XLVIII, pt. II (1935), 173.
[4] Once again, specific exception must be made of the areas which are above the winter snow-line. These areas – 2,000 square miles in New South Wales and Victoria, 2,500 square miles in Tasmania – are only 0.1 per cent of the total area of Australia.

agreed explanation has been as yet forthcoming; but some people have sought an explanation in the contrasted use of fire made by the Aborigines and by their European supplanters.

The Aborigines, it has been argued, kept their hunting grounds 'open' by burning them regularly and lightly; in doing this they had nothing to lose, because they possessed no more property than they could carry with them on their travels. By contrast, the Europeans had valuable property to protect: fire was therefore their enemy, except when they used it with intent to procure shorter and sweeter pasture for their grazing animals. If they had used fire in the Aboriginal fashion, the argument concludes, they would have had no need to lament the invasion of their properties by useless scrub.

In conformity with this reasoning, some Australian anthropologists have joined the school of thought which sees the world's grasslands, not as a natural 'climax', but as man-made.[1] Australian botanists would nowadays agree that the human, or 'anthropogenic', factor must be given its due place – whatever that may be – along with the 'natural' factors.[2] One botanist, W. D. Jackson, gives it the dominant place. His researches in Tasmania have made him a strong supporter of the argument reported above.[3]

Tasmania, however, is a long way from Monaro. Moreover, the last of the Tasmanian Aborigines had been dead for three-quarters of a century before Jackson began his field research. We need to get much closer than that to our own place, and to Aboriginal time in that place. The closest we can get is Gippsland, from the 1860s to the 1890s, when A. W. Howitt was police magistrate there. By and large, we may count ourselves fortunate. After all, the eastern district of Gippsland had been an integral part of the Squattage District of Monaro. Moreover, Howitt knew the Aborigines of Gippsland as persons and made an outstanding contribution to the scientific study of their society. He also knew the country and its vegetation, almost as well as he knew the people. In 1890 he published an important botanical paper, *The Eucalypts of Gippsland*.[4] In this paper he made two firm statements: first, that the Aborigines had burnt the country every year and thereby kept it open and grassy; secondly, that the white settlers protected their property against fire and thereby encouraged the return of trees and scrub. In support of the second state-

[1] See particularly Norman B. Tindale, 'Ecology of Primitive Man in Australia', in *Biogeography and Ecology in Australia* VIII (1939), and Rhys Jones, 'The Geographical Background of the Arrival of Man in Australia and Tasmania', in *Arch. in Oceania* III, no. 3 (1968).
[2] See C. W. E. Moore, 'Distribution of Grasslands', in *Australian Grasslands*, ed. R. Milton Moore (A.N.U. Press, 1970).
[3] See *Atlas of Tasmania* (Hobart, 1956), pp. 30–5.
[4] *Trans. Roy. Soc. Vic.* II (1890), 83–120. The paper is beautifully illustrated with drawings made by Howitt's daughter, Annie.

ment, he cited his own observations throughout three decades in seven representative districts. He then continued:

> I might go on giving many instances of the growth of Eucalyptus forests within the last quarter of a century, but those I have given will serve to show how widespread the re-foresting of the country has been since the time when the white man appeared in Gippsland, and dispossessed the Aboriginal occupiers, to whom we owe more than is generally surmised for having unintentionally prepared it, by their burnings, for our occupation.

Howitt's evidence might seem to close the argument, so far as Gippsland is concerned; but the opposite is true. In a recent paper on 'Bushfire Frequency and Vegetational Change in South-eastern Australian Forests',[1] N. A. Wakefield has argued that the white people caused the return of the scrub, not by stopping fires, but by starting them. Wakefield is a well qualified inquirer, for he was trained as a biologist and thereafter made himself an expert in the history of exploration and settlement in East Gippsland, where he was continuously resident from the 1920s to the early 1950s. His paper opens with the quotation of a long statement by a grazier, K. C. Rogers, whose direct acquaintance with East Gippsland stretches back from the late 1960s to the early 1900s, when his father acquired a large property on the Wulgulmerang plateau.

> Over a period of years before we came to the district, it had been the accepted practice to burn the bush, to provide a new growth of shorter sweet food for the cattle.
> The practice was to burn the country as often as possible, which would be every three or four years according to conditions. One went burning in the hottest and driest weather in January and February, so that the fire would be as fierce as possible and thus make a clean burn ... We would light along the rivers and creeks, so that the fire would roar up the steep slopes on either side, making a terrific inferno and sweeping all before it. The hotter the fire, the sweeter and better the food for the cattle after the new growth came. The tablelands received special attention ...

Rogers goes on to say that the sweet grass grew coarse again after a season or two and that the scrubby growths proliferated. His evidence has significance for the forestry experts who are now practising a policy of 'controlled burning', as the most effective means that they can see of preventing uncontrollable and ruinous bushfires.[2] Wakefield consulted some of these experts and held supplementary discussions with Rogers. In principle, if not in every detail, he has accepted Rogers' evidence. His

[1] *Vict. Nat.* LXXVII (1970), 152–7.
[2] The case for controlled burning has been ably stated in successive papers by A. S. McArthur of the Commonwealth Forestry and Timber Bureau, R. G. Vines of C.S.I.R.O., W. R. Hatch and W. R. Wallace of the Forests Department of Western Australia. Recently, the case has been carefully examined in the Australian Conservation Foundation's Viewpoint Series No. 5, *Bushfire Control and Conservation* (1970). See p. 179 below.

own conclusion is 'that fires due to lightning and aborigines were collectively less frequent than those during the era of colonisation'.

Howitt's argument – although Wakefield does not directly cite it – has thus been contradicted. Nevertheless, the contradiction may not be quite so stark as appears at first sight. To begin with, three comments need to be made on the Rogers statement.

1. It has reference only to a limited area of dry sclerophyll forest country on a plateau upwards of 3,000 feet in height. Howitt knew this country well, for it was on his route to Tubbutt, the home of his friend William Whittakers;[1] but he did not include it among his seven case studies.

2. Rogers gives evidence at first hand on the practices of the graziers from the early 1900s onwards; but Howitt's evidence, submitted in 1890, reaches back to the mid-nineteenth century. It still remains to be proved that the frequent, fierce burning described by Rogers was practised in the earlier period.

3. Rogers makes no statement at all about burning by the Aborigines. He was in no position to make one, because the few surviving Aborigines of his district had become displaced persons on a distant mission station many years before he was born.

Far more research – by ecologists and foresters, prehistorians and historians – will need to be forthcoming, before these controversies about the use of fire by black people and white people can be resolved. By present indications, the prospect of resolving them looks more promising in Gippsland than it does in Monaro. There, natural scientists seem still to agree with R. H. Cambage that the explanation of the treeless plains and park-like savannahs is to be found chiefly in the composition of rock and soil. Historical research in Monaro has not as yet got very far. Prehistorical research is only now getting started.

The research will need to be both meticulous in its local detail and intelligently attentive to the broad perspectives that have been revealed by the study of grassland and savannah elsewhere in the world. We may have to wait quite a long time before the large gaps in our knowledge begin to be filled. Meanwhile, we may not casually dismiss Howitt's idea that the first discoverers of Monaro had spent many thousands of years in so changing the surface of the land as to make it easy for their white successors to oust them within a single decade.

A note on 'whites' and 'blacks'

These terms of description found early and firm lodgment in Austral English. They are used here for that reason only. Neither word correctly describes the skin colour of the people to whom it refers. It would be closer to the truth to call Europeans pink and Aborigines brown.

[1] See p. 38 below.

PART II

NEW ARRIVALS

1. Squatting

During the 1830s and 1840s the word 'squatter', as used in Australia, rapidly changed its meaning. In 1835, rich and powerful persons were still using it as a term of opprobrium for the lowest type of the undeserving poor; ten years later, they were using it as a term of respect for members of their own deserving class. This lexicographical revolution has been sufficiently studied and explained; but argument continues about the historical significance of the squatters.[1] Were they liberators, or were they oppressors? What, specifically, were their attitudes to convict labour and free immigration? Were they the innocent victims of booms and slumps or were they in some degree the makers of them? From what sources did they get their capital? What part did they play in the political fights of the 1840s? What was their character as a social class? What connections did they have with other classes? Questions such as these are so interesting that we shall find it hard to push them aside; but this we must try to do, if we are to make headway with our own inquiry into land use in Monaro.

It seems a sensible idea to make personal acquaintance with some of the squatters. Let us look at them, to start with, as a novelist has portrayed them; after that, we can look at the self-portraits which some of them painted, quite unselfconsciously, in their letters and diaries. The novelist is Henry Kingsley, who spent four years in Australia during the 1850s and fell in love with the country, the more so as he did not have to stay in it for ever. His *Geoffrey Hamlyn* introduces us in its opening chapters to the Buckleys and the Brentwoods, two English families conspicuous both in county society and in loyal service to king and country, but alarmingly on the down-grade financially. Lower in the social scale, but within the same circle of friendship, we meet the widowed parish priest, John Thornton, and his beautiful daughter, Mary. Three honest country gentlemen – James Stockbridge, Thomas Troubridge and the narrator, Geoffrey Hamlyn – love that girl; but she has a mad infatuation for the villainous George Hawker. Moving upwards again in the social scale, we meet a Prussian polymath of noble birth, Dr Mulhaus, who has made his home with the Buckleys. We also meet Frank Maberly, an Etonian and Cam-

[1] See e.g. S. H. Roberts, *The Squatting Age in Australia, 1835–1847* (M.U.P., 1935 and 1965); and D. W. A. Baker, 'The Squatting Age in Australia', a review article in *B.A.H.* v, No. 2 (August 1965).

bridge man, who has won the highest academic and athletic honours before taking service in the army of Christ; he joins Devonian society by vaulting a high gate after running four miles up hill in clerical broadcloth and heavy boots; his timing of the run – 21 minutes – vexes him, until he reminds himself how steep the last mile has been.

Before long, we must meet these people again in the Squattage District of Monaro.[1] Various impulses have propelled them thither. Major Buckley has sold his shrunken estate to the local brewer and faces a precarious future: 'I have only ten thousand pounds,' he calculates, 'and how am I to bring up a family on the interest of that?...I am thinking of emigrating. To New South Wales.' Captain Brentwood, we may assume, has made similar calculations. George Hawker has miscalculated, and is now serving a life sentence in Van Diemen's Land. Mary, his ill-used wife, has decided to make a new life in New South Wales for her little son and her maiden aunt and herself; for the time being, she is with the Buckleys. Close at hand are two of Mary's rejected lovers; the third is on his way to join them. Dr Mulhaus arrives and stays on as tutor to young Sam Buckley. Frank Maberley arrives in the dress of a Colonial Dean. 'Half Devonshire,' exclaims Mary, 'is here.' And here the Lord blesses them. Within no time, so it seems, Major Buckley has 28,000 sheep, a fine new house with grape and passion vines festooning its verandahs, a flower garden ablaze with geraniums and petunias, racehorses in the stable, claret in the cellars, money in the bank. The Buckleys now dine late in the evening, as they used to do in England. Sam Buckley, carefully segregated as a boy from the convict servants, is growing up a young colonial gentleman. He falls in love with sweet Alice Brentford. Her brother James is an applicant for the king's commission in India.

But trouble is afoot. Frank Maberly goes to a shepherd's hut and preaches on the text, 'Servants, obey your masters'. His sermon is homely, plain, sensible and interesting; but not all the servants hear him gladly. Bad men are lurking in the mountains. Their leader is the escaped convict, George Hawker. Mary Hawker hears the news and is terrified. Captain Desborough of the Border Police organises a force to round up the ruffians. In the battle that ensues, George Hawker shoots his own son.

Peace follows the storm. George Hawker goes to the gallows. Frank Maberly goes to Patagonia to preach the gospel and win the martyr's crown. James Brentwood wins glory in Afghanistan and the Crimea. Almost everybody else goes 'Home'. Captain Desborough inherits an Irish earldom with an income of twenty thousand pounds a year. Dr Mulhaus is recalled by the King of Prussia to take office as Prime Minister. Most interesting of all is the return of the Buckleys and the Brent-

---

[1] Novelists are at liberty to make their own landscapes; but the descriptions in *Geoffrey Hamlyn* point to the neighbourhood of Tubbutt. (See p. 38, below.)

woods. The young people arrange it. On the eve of their marriage, Sam tells Alice what he has in mind:

> Alice, I have had one object before me since my boyhood . . . I want to buy back the acres of my forefathers. I wish, I intend, that another Buckley shall be the Master of Clere, and that you shall be his wife . . .
>
> Think of you and I [*sic*] taking the place we are entitled to in the splendid society of that noble island . . . What honours, what society has this little colony to give, compared to those that open to a fourth-rate gentleman in England? I want to be a real Englishman, not half a one . . . I don't want to be young Sam Buckley of Baroona. I want to be the Buckley of Clere. Is not this a noble ambition?
>
> 'My whole soul goes with you, Sam,' said Alice. 'My heart and soul. Let us consult, and see how this is to be done.'

They consult. Sam buys sheep country in northern New South Wales and building lots in Melbourne. By the mid-1850s, he has 118,000 sheep on his new runs and is making 1,000 per cent profit on his Melbourne speculations. He is by now one of the richest of Her Majesty's subjects in the Southern Hemisphere. His dream comes true. He returns home, buys Clere, and for good measure discovers himself to be the inheritor of Beaulieu Castle. In that majestic neighbourhood the returning squatters gather. Only Mary Hawker and Tom Troubridge stay put. They marry. Their second child, a son, grows up to become one of the colony's finest riders.

In Henry Kingsley's novel, false sentiment does not invariably smother truth. Kingsley has a blind eye for hard times, but a clear eye for great expectations. An emigrant Scot, David Waugh, expresses a comparably exuberant optimism in his letters home during the mid-1830s. Like many young men of that time, Waugh makes his colonial start by managing another man's property. His employer allows him to run stock of his own; a neighbouring property becomes available to him on a five-year lease; there is vacant Crown Land not too far away. In New South Wales, he says, anybody can get land for next to nothing; all a man needs is a few hundred pounds for the purchase of cattle and sheep; their natural increase will do the rest. He produces figures to prove that those few hundred pounds will soon be bringing in an income of £1,500 per annum. Naturally, a man has to work hard and he may run into bad luck; but there can be no doubt at all of his getting rich. The only thing that he cannot predict is how long it will take him to get rich.

Waugh's letters found a publisher and an enthusiastic reading public. A sardonic squatter declared later on that David Lindsay Waugh had gulled half England and Scotland. If this was true, the author himself should be counted with the other victims. He left his anchorage in the Goulburn district, packed his dray and tarpaulins and drove his cattle and sheep southwards into 'the new country', until he came to the Devil's River. The land he took there lay to the west of the Monaro Squattage District, but well within the territory of the Bogong moths. Before long, he had good reason for cursing the day when he squatted on it. The blacks nearby killed two of his shepherds, took hundreds of his sheep, ransacked his house and left him without even a pannikin or a pair of socks. Soon after that disaster, some travelling sheep infected his decimated flocks with catarrh. Devil's River was living up to its name. Waugh fled the place and told everybody who would listen, including the Governor, that he was ruined for life. Yet we meet him again, about ten years later, in a pleasant coastal town of the Illawara district, a prosperous and respected citizen, a pillar of the kirk and a promoter of all good causes.[1]

Another young Scot, Farquhar McKenzie, was recording in his diary during the late 1830s hopes, fears and griefs of a different kind.[2] He was the son of Captain Kenneth McKenzie, from Kerrisdale near Gairloch in Ross-shire. He had the Gaelic. He was clannish, pious and perpetually homesick. He cherished no dreams of getting rich quick but was sober and prudent in his business dealings. At the end of 1836 he landed in Sydney with approximately £2,000 – not all his own money – and with the intention of buying sheep and travelling with them to 'unappropriated lands'. In a reconnaissance up-country, he met a friend of his family, an ex-officer named Murchison, now a landowner at Tarradale near Goulburn. He and Murchison made a business agreement and on 30 January 1837 he set off for Monaro with 2,450 sheep, of which the greater number belonged to his partner. On 19 February he wrote in his diary:

> I have now pitched my tent on Monaro in a beautiful little valley which I have named Kerrisdale[3] . . . My party consists of myself, an overseer named William Bell and 8 of Mr Murchison's assigned servants – they are of course all convicts but with one or two exceptions seem very good men.

They were not, however, good company. Farquhar McKenzie had to fight a never-ending battle with his own almost insupportable loneliness. To be sure, four or five miles away he had a neighbour, Mr Coghill, whom he thought 'a decent man, being a native of Caithness'. Mr Coghill's son, in his strange Australian way, seemed equally decent:

> his son (though perhaps born in Scotland) is a fair specimen of a country born (currency) lad – of fair complexion, tall and raw-boned – rather reserved and

---

1 See D. Waugh, *Three Years' Practical Experience in New South Wales, being Extracts from letters to his friends in Edinburgh from 1834 to 1837* (Edinburgh, 1838); T. F. Bride (ed.), *Letters from Victorian Pioneers* (Govt. Printer, Melbourne, 1899), nos. 34, 47; *The Petition of David Lindsay Waugh to H.E. Sir George Gipps*, 9 October 1830 (C.O. 201/309); *Illawara Mercury* (2 September 1879, Waugh's obituary). Devil's River is now always called Delatite River.

2 The Journal of Farquhar McKenzie was presented to the Mitchell Library in 1941 by his son D. McKenzie.

3 Kerrisdale is in the Nimmitabel district, but it was never marked on any map.

silent – knowing and caring for nothing but sheep and cattle – Good Runs – Native Dogs – Slabs, Bark hurdles, stock yards and Paddocks, but probably sober and steady, good tempered and hospitable.

At a different level of the pastoral society, Farquhar McKenzie recognised the same Australian stamp of character.

> Wandering 'Stockmen' (as Cowherds are called in this country) often come to light their pipes at my fire or to have a yarn and pot of tea with the men – these fellows are generally mounted on a poor looking ewe necked horse – which is however handy, well trained and sure footed – the rider's heels are armed with a formidable pair of spurs and in his hand he carries a whip with a short handle but a lash at least 3 yards in length used for driving cattle – his talk is of Rum, Tobacco, Cattle, Horses, Brands, Increase and Stockyards, always interlarded with abundance of oaths and imprecations . . . He prides himself on knowing all the ranges, creeks, gullies and swamps for 100 miles around – the situations and distances of different places and where particular cattle and horses are to be found.

Here we have a portrait of the Man from Snowy River, painted half a century before Banjo Paterson made a folk hero of him.

The exiled Scot recognised this new breed of men; he appreciated its good qualities; but he did not find the breed companionable. Throughout the hours of daylight, except for a break at midday for thinking and writing, he fended off his loneliness by keeping himself at work; but he could not fend it off after the sun went down.

> The evenings [he wrote on 23 June 1837] are now very long, there being nearly 14 hours of darkness and I am badly off for Books and still worse for companions, not having one – my greatest comfort is the hope of Hector [his dearly loved brother] coming down to this country and of our, one day, revisiting our native land . . .

Often, when he went to bed, he had nostalgic dreams of his family and friends in Scotland. For example:

> Last night I dreamt I was dancing Quadrilles with Misses A. and K. Mck., Mary playing the piano for us . . .

He tried to keep his loneliness at bay by concentrating his mind on the truths of the Christian religion, by making little sketches of things seen or remembered, by writing down scraps of miscellaneous information and reflection that were stored in his head, or perhaps collected in a book he had kept with him. He filled some pages of his journal with verses by Mrs Hemens and other favourite authors. He wrote verses of his own.

> But when at last my spirit flies
> Some convict's hand will close these eyes
> And this poor clay perchance be laid
> Beneath yon gloomy Banxia's shade

No friendly eye shall drop a tear
No stone to mark the spot appear
And no lament swell in the gale
Save the discordant wild dog's wail

The native shall avoid the glade
Fearful to meet the white man's shade
And o'er my cold and narrow home
The Kangaroo and Emu roam.

At the end of his sad song, he consoled himself with a vision of the everlasting life and love awaiting him in the heavenly realms. Yet he was only twenty-five years old and had still ahead of him thirty-eight more years of life on earth. The years of slump in the early 1840s proved to be the worst. In December 1938 he said farewell to Kerrisdale in Monaro, overlanded with all his animals to the Port Philip District, and established a second Kerrisdale in the valley of the King Parrot Creek. There, for two or three years, his worldly affairs went reasonably well; but in July 1841 he suddenly stopped writing in his diary. On 22 January 1843 he opened it again and wrote:

> It is now long since I last opened this record of my monotonous and uninteresting life. Since last date I have suffered a good deal in mind, not so much from having ruined myself as from having brought distress on others from my credulity and inconsiderate conduct.
>
> I am pennyless but still have to thank God for many mercies – especially for having given me kind friends in those whom I have so greatly injured.

He had in mind the family of his partner Murchison. Working together, he and Murchison restored their fortunes as the colony climbed out of the slump. In October 1846 he married Patty Murchison. He was beginning to enjoy such contentment and happiness as he had never expected to find in this world. Homesickness for the Scottish Kerrisdale had long since ceased to gnaw at him. In his journal, on the rare occasions when he still opened it, he jotted down his observations of Australian plants and fungi. He ended his working life as Chief Inspector of Stock in Victoria.

The race of sturdy pioneers is variously constituted. We have met two temperamentally opposite Scots; let us now meet two Englishmen. The first of them, Robert Dawson, had read David Waugh's book, or something of that sort, and in a letter from Monaro in December 1842 denounced it as a fraud.

> With regard to this wretched and miserable colony I would never advise any person to come out here that can possibly stay at home, for they will meet with nothing but disappointment . . .
>
> Should you, my dear friend, know any one that is likely to come out here, let them read this letter, as there are hundreds that would go back, if possible, if they were sure of going to the workhouse . . .

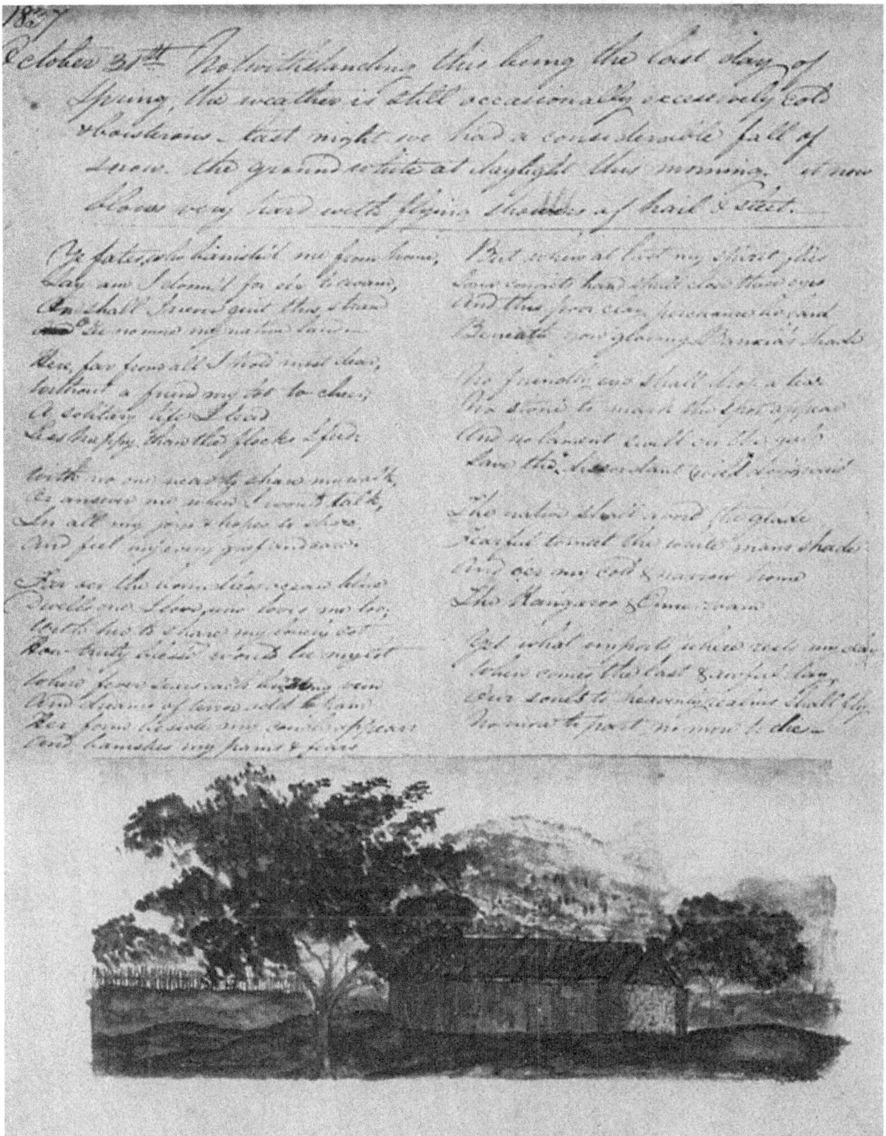

I   A page from the diary of Farquhar McKenzie, 1837

II   Tombong, the Whittakers' first home, early 1840s

III Teamster's camp near Cooma, early 1900s

IV Crossing the Snowy River, early 1900s

Squatting

All the books that have been written upon the colony contain nothing but falsehood from beginning to end, and it is only to bring people out, as they well know when they are once here they cannot get back, and then they are caught in the trap . . .

For this embitterment Dawson had sufficient reason. He had landed in Sydney in 1839, during the worst drought that anybody could remember. The drought lasted until 1842, by which time the colony was suffering the worst slump that anybody could remember. Dawson had brought out with him all his capital, approximately £2,000. Three years later he had lost it all, or nearly all. In 1839 he had bought cattle at £4 a head; next year they fell to £2 a head, and year by year the fall continued. For three successive years he had sown 10 acres of his run with wheat, but only once had he reaped a harvest. He was living with his wife and sons in a small hut with a dirt floor, cracks in the walls through which the wind whistled; but no windows. An English stable, he said, would be by comparison a palace. His neighbours, as he saw them, were all drunkards and thieves. The nearest clergyman was 70 miles away. 'I would go home to-morrow,' he declared, 'if I could get £50 to pay my passage.' Yet he had to admit that his wife, notwithstanding her sufferings, did not want to go home. We may guess that she was looking forward to better times for her children, and may hope that she lived long enough to see Robert Dawson Junior a prosperous citizen of Cooma and its first police magistrate.[1]

Our second Englishman, William Whittakers, arrived in the colony only two months after Dawson; but he made the opposite response to the challenges of drought and slump.[2] Looking for the reason why, we can discover no conspicuous advantages of birth and education that Whittakers possessed; nor have we any ground for believing that he had brought with him more money than Dawson had. Youth was his great advantage. Whereas poor Dawson, lamenting his 'latter end', must have been in the middle or late fifties, Whittakers was only twenty-five years old. He had zest for action and within a few weeks paid £568 for a cattle run on the Snowy River; yet he was also sufficiently prudent to set a limit to his commitments. He had no dependents to provide for and consequently no need to stock his run heavily. Still, we must assume that he hired a manager, for very soon we see him making frequent visits to Sydney. There he courted Louisa Grant. His marriage to her in November 1841 was his great stroke of fortune.

[1] Letter from Robert Dawson to George Arnatt, Esq., 10 December 1842 (printed copy in the N.L.A.). Dawson's run, 'Gellimatong', was close to Cooma on the west and was gazetted in 1848 in the name of the younger Robert, on whose career see *Back to Cooma* (1926), p. 60.

[2] See *The Whittakers Story*, compiled by Clyde M. Whittakers (typescript 1956–60), and *William Whittakers' Day Book* (contemporary manuscript). Both are in the N.L.A. See also Mary H. Walker, 'Pioneering Pageant', *The Bulletin*, 20 March 1953. William Whittakers' father was a merchant draper and an alderman of Chester.

Her father, an army captain, had died in the mid-1830s and an aunt had brought her out to live with her mother's people, the Moores – a family well established in colonial employment and on colonial land.[1] At the age of seventeen, she was enjoying the parties of the governing clique in Sydney; until she met William Whittakers, a life in the bush was the last thing she would have chosen for herself. 'You know, dearest William,' she wrote to him, 'I never wanted to be a settler's wife.' But no settler ever had a better one. She had little or no money to put into the common stock; but her family connections proved useful; by a mutually profitable arrangement, the young people went to live with her Uncle Thomas on Burnima, a very good property, not too far distant from William's cattle station. There Louisa had the first of her eleven children[2] and there she served her hard appenticeship in bush housekeeping. The only thing that vexed her was her husband's tolerant attitude towards her uncle, whom she detested; but in 1846 the great day came when she and William moved into their own home, a slab hut with a bark roof and dirt floor and no glass in the windows. It was not on the original Whittakers station but on Tombong, to the west of the Delegate River and about 15 miles north of the recently surveyed Victorian border. Four years later, they moved to Tubbutt, about the same distance south of the border. Henry Kingsley must surely have seen Tubbutt, for it was precisely the setting in which Major Buckley made his elegant home. The Whittakers lived at Tubbutt for twenty-two years and made a good home there – if not an elegant one.

These references to Tombong and Tubbutt would be superfluous, were it not for the daybook or diary that William Whittakers kept in both places. With some intermissions, it faithfully records the daily round of life on a station – predominantly a cattle station – from the 1840s to the 1860s. For historical students of various kinds it is a magnificent record. Later on, we shall make occasional use of it; for the present, let us simply take note of its style. The diarist is invariably laconic. He calls himself W.W. and his wife Mrs W. He makes entries such as the following: 'Mrs W. confined. Jamie carting manure.'[3] His recording of his own doings is equally matter-of-fact; but its cumulative effect is to make us fear that W.W. may be killing himself with overwork. His mother, as she read his letters, had the same fear. It broke her heart, she told him, to think of her delicate boy working like a galley slave, and poor Louisa living in *such slavery*, and their sons growing up like common servant boys. She reminded him of the easier life that farmers lived at home. She persuaded his brother-in-law, a retired wine merchant of Chester, to offer him a perpetual, interest-

---

[1] See *A.D.B.*, vol. 2, entry on Moore, William Henry, in which is included some reference to his brother, Thomas Matthews Moore.
[2] Nine of her children survived. One child, to her great distress, died unbaptised.
[3] One of the entries for 8 April 1866.

free loan that would set him up as a farmer in North Wales or Ireland. All to no purpose. Her son and his wife were in love with their hard life. Years later Louisa Whittakers exclaimed, when somebody told her that one of her sons had been thinking of leaving his home country – 'It would have broken my heart.'

In their declining years, the Whittakers lived less strenuously on a property they had bought on the coastal plain of Gippsland. They still held Crown leases in the high country; but their sons were now the rough riders. They still remained in the cattle business; but now they were fattening the stores, rather than raising them. William died in 1893; Louisa in 1910. The Whittakers' estate, to be divided among the nine surviving children, was valued at approximately £80,000. That sum fell a long way short of Sam Buckley's quickly got fortune; but it was a fair return on sixty years of hard work. The best return, of course, was the Whittakers' clan – the nine children with their wives and husbands and sons and daughters.

Let us now meet two individuals, Stewart Ryrie and W. A. Brodribb, who were well acclimatised to Australia before they squatted in Monaro. We need to remind ourselves that the long leap from the British Isles was not the normal manner of entry; infiltration from settled districts close at hand occurred more frequently. Conspicuous among the infiltrators were men of means – officials, naval and military officers, medical men, shipping men, merchants – who had received grants of land during the early decades but needed out-stations to carry the increase of their livestock. Some of these people went themselves to their out-stations; some of them sent managers; some of them sent their sons. Commissary-General Ryrie, of Arnprior near Braidwood, sent three sons. Or perhaps the sons set forth on their own account, to establish themselves, then or later, on stations of their own.[1]

Stewart Ryrie, with whom we are concerned, began as a manager at Coolringdon before establishing himself at Jindabyne. During that interval he had made himself uniquely acquainted with the geography of Monaro. He did this in 1840, after receiving an official commission to traverse and report upon the entire Squattage District – the main landmarks, as determined by 'celestial observation'; the mountain ranges, with their links and spurs and the lie of the land between them; the flowing streams, with their depth and breadth and crossing places; the Aborigines, with an estimate of their numerical strength and their disposition towards white settlers; the pastoral stations already established, and the opportunities for establishing more; the most promising sites for towns. For a man who possessed no training in any science and was no more than an amateur surveyor, this was a difficult assignment; but Ryrie acquitted

[1] *Back to Cooma*, p. 69.

himself well. He criss-crossed the country systematically, took careful notes, and made simple profile sketches. His report remains worth reading, even today, because it tells us what a practical man saw on the surface of the land 130 years ago. That, precisely, is what we want to know: for example, we want to know, and the report tells us, what the alpine watersheds looked like at the time when drought-afflicted squatters were barely beginning to use them for summer grazing.[1]

In contrast with Stewart Ryrie, W. A. Brodribb possessed intellectual depth. Starting with action and observation at the surface, he gradually achieved an understanding of all the problems – technological, financial, political – of the pastoral industry of his time. Throughout his working life he kept precise records of all his enterprises and in his old age he drew upon them to produce a memorable book.[2] His formative years – nine to twenty-four – had been spent in Tasmania, at a time when progressive flockmasters there were as good as any in New South Wales. Tasmanian pastures, however, were becoming too cramped for the expanding flocks and herds; consequently, the 1830s witnessed an exodus of enterprising Tasmanians to the mainland. Most of them went to the newly opened Port Philip District; but a few, Brodribb among them, went to the mother colony. In 1835 he landed in Sydney with a capital of £500. He immediately spent £15 of that small sum in buying a pony, saddle and bridle and a pair of saddle bags; thus equipped, he set out on an expedition to Monaro. At the cost of £6, he covered many thousands of miles, made a systematic inspection of the district, and located on the upper Murrumbidgee an unoccupied area of about 100 square miles. In association with a friend better supplied than he was with capital, he squatted on this land with cattle. The venture proved profitable. From that time onwards, Brodribb possessed the financial means to launch out on his own.

Between 1836 and 1842 he launched out in various directions – in the valley of the Murrumbidgee, running sheep; in a cattle and sheep venture near Goulburn; on stations both north and south of the Murray. Like a South African trekker, he moved about with his sheep and cattle to wherever he could find grass and water. Thus he survived the drought. In surviving it, he gathered experience which he turned later on to good account as an innovating practitioner and evangelist of water storage. But he was not so competent or fortunate in coping with the slump. After overlanding his animals to the falling Melbourne market, he held them back for a rise that did not come. Then he joined a syndicate which proposed to open a sea route to pastoral Gippsland. Technically, that venture

---

[1] Stewart Ryrie, *Journal of a Certain Tour in 1840* (M.L.). On the beginnings of alpine grazing, see pp. 132–6 below.

[2] *Recollections of an Australian Squatter, or Leaves from my Journal since 1835*, by Hon. W. A. Brodribb (Sydney, 1883). See also entry on Brodribb in *A.D.B.*, vol. 1.

was a success; financially, it was a disaster. Brodribb wrote off his losses to experience; never again, he vowed, would he speculate beyond his means. He returned to Monaro to become chief manager there for a superb employer, William Bradley. In 1855, after serving Bradley for twelve years, he drove his own cattle and sheep over the Australian Alps to his own station in the Riverina – to Wanganella, a name destined for fame in the annals of the Australian pastoral industry.

Very few squatters were so articulate as the six individuals whose portraits we have been studying. Although our inquiry is only incidentally sociological, we must now take a broader view of the squatting society. Such a view emerges clearly from the letters of John Lambie, Commissioner of Crown lands in Monaro, to the Colonial Secretary in Sydney. Like the Commissioners in the other Squattage Districts, Lambie took office in 1837, following the Act passed the previous year by the Legislative Council, 'for obtaining such recognition of the title of the Crown from all occupiers of waste lands as will prevent any difficulty in their future resumption by ordinary legal process'. The Act embodied a compromise: on the one hand, it declared emphatically that the squatter was *not* the owner; on the other hand, it gave the squatter legal status as a tenant of the Crown from one year to the next, provided that he was a man of good character and paid every year a fee of £10 for his licence to depasture stock. It was the Commissioner's duty to see that these conditions were fulfilled; but, to begin with, the Commissioner possessed inadequate means for performing that duty. An Act of 1839 in some degree made the deficiency good: it gave the Commissioner wide-ranging judicial powers within his district, and gave him a small police force (composed, to begin with, of convicts) for the enforcement of his decisions and for the maintenance of public order. For the financing of his skeleton establishment, a tax, additional to the licence fee, was levied on all stock within the Squattage District – one penny for each sheep, threepence for each head of horned cattle, sixpence for each horse.[1]

The Commissioner had laid upon him a preliminary and basic duty: to perambulate his district, record his itinerary and collect the materials for a census. For this purpose he was provided from the beginning with forms which remained unchanged throughout the following decade. On

---

[1] The writer has decided to keep clear of the controversies about colonial land policy. Historians nowadays are paying rather less attention to their ideological aspects, and rather more attention to the problems of land survey. For example, the boundary of the Nineteen Counties, as defined by Governor Darling in 1829, used to be depicted as a kind of Chinese wall; but the very next year an additional county (see map on p. 7, above) was gazetted. The progress of survey, as enjoined by Lord Bathurst in 1825 (see p. 8, n. 5, above) was never stopped. However, the understaffed Department of the Surveyor-General was quite unable during the 1830s to keep pace with the rapid territorial expansion of the pastoral industry. Crown land that could not be surveyed could not be alienated by any legal process. Hence the much used phrase, 'the limits of location'.

these forms the Commissioner recorded in ruled columns the following information: the extent of each day's journey, the name of each station visited, the name of the licensee, the name of the person superintending, the number of residents, the number and nature of the buildings, the number of acres in cultivation, the numbers of cattle, horses and sheep. The Monaro Squattage District, be it remembered, was immense; in 1840 Stewart Ryrie spent six months at least in traversing it. When we read Lambie's itineraries and letters, we are surprised at his success in covering so much ground; but we should be still more surprised if a good many people did not manage from time to time to escape from him with their livestock into the bush. Even so, Lambie gathered sufficient information to make it hazardous for any squatter to evade his obligation to apply for a licence. Naturally, it took some time for this knowledge to sink in; during the early years, Lambie had to spend a good deal of his time in reminding the squatters that government was, at long last, catching up with them.

Lambie did not collect the fees; nor did he, in the early years, issue the application forms. Under the Act of 1836, the administrative headquarters of the Squattage District were not in Cooma, but in the magistrate's office at Braidwood.[1] Thither the squatter had to ride to get his application form. Having got it, he had to sign it and bring or send it to the Commissioner in Cooma. There, Lambie had to make up his mind whether or not he would certify that the applicant was a fit person to receive a licence. If he did so certify, the applicant or his agent had to pay the fee and receive the licence at the Colonial Secretary's office in Sydney. Lists were kept there of the persons licensed to depasture stock in Squattage Districts. From time to time, these lists were published in the *Government Gazette*. Tracking defaulters was the Commissioner's business; fining them was the business of the Board of Magistrates in Braidwood or (later on) Queanbeyan. Lambie invariably recommended the full fine, which meant that a year's tenancy of a run cost the defaulting squatter £20 instead of £10. In the 1840s, when Lambie himself possessed jurisdiction in Cooma, his severity drew an occasional remonstrance from the officials in Sydney. They failed to shake his conviction that maximum severity and maximum respect for the law were the two sides of the same coin.[2]

Before long, defaulters on the licence fee were only a small minority. Consequently, the larger squatters had more time and energy to spare for lodging complaints against the smaller squatters. They pursued with particular rage the men whom they suspected of demoralising their station hands by getting them drunk on spirits and inveigling them into

[1] Late in 1838, following a request from Lambie, the headquarters were shifted to Queanbeyan.
[2] Lambie/C.S.O.: 10 December 1846. [From now on, references thus abbreviated will identify Lambie's letter to the Colonial Secretary.]

the traffic in stolen stock. Lambie carefully investigated every complaint and quite often sought advice from the officials in Sydney: in his letters, we are able to see the common man in Monaro fighting for a foothold in cracks and crannies of the squatting society. Some common men were as disreputable as the respectable people said they were – sly grog sellers, stock thieves, traffickers in stolen stock. Whether or not they were applicants for a licence made no difference; Lambie sent them to the magistrates for quick trial and banishment; after the Act of 1839, he tried and banished them himself. Quite often, however, he decided that the respectable had no justification for their complaints; that a good blacksmith, for example, was the kind of man Monaro needed, even if he sometimes sold a few pannikins of rum. His decisions were invariably well considered and fair-minded. On a case that he referred to Sydney in September 1838 an official minuted: 'This is rather a doubtful case, but from the very careful manner in which this Commissioner performs his duties, it may be safely left to him to deal with Haydon according to his own discretion.'[1]

He was so anxious to attract good craftsmen to Monaro that he wanted them to be granted exemption from the licensing system; or, failing that, he wanted them to be charged at a lower rate.[2] He showed the same realism – and, one may add, the same humanity – in support of the attempts which some poor people were making to establish themselves in a small way as graziers. Ticket-of-leave men, for example, sometimes received sheep and cattle in lieu of money wages; Lambie considered it anomalous that they could not get legal tenancy of land on which to graze their animals.[3] He fought a persistent battle for justice to the widow of a ticket-of-leave man who had been granted a conditional pardon, but had died before the notification of it arrived; the horses which the husband had possessed, he argued, belonged in equity to his wife.[4] He wanted just as much to see justice tempered with mercy towards people of a higher position who, through no fault of their own, had fallen on evil times: for example, towards Ellen Woodhouse, who had been left a widow with eight young children; she had stock and a station, but not enough money to pay for the depasturing licence. Lambie strongly supported her petition to Governor Sir Charles Fitz Roy for remission of the fee. The petition failed. Nevertheless, Ellen Woodhouse battled her way through the hard times.[5]

[1] Lambie/C.S.O.: 2 May, 13 and 21 June, 15 August, 21 December 1837; 10 April, 29 May, 23 July, 9 and 25 August, 4 September, 25 October, 6 December 1838 – and similarly in the following years.
[2] Lambie/C.S.O.: 16 August 1839.
[3] Lambie/C.S.O.: 15 May 1845.
[4] Lambie/C.S.O.: 6 November 1846 and enclosure 1 June 1841.
[5] Lambie/C.S.O.: 20 November 1846. In 1848 (see p. 52, below) Ellen Woodhouse was lessee of Inchbyra. Today, Gegedzerick is Woodhouse property.

Towards the mid-1840s, Lambie found himself increasingly called upon to settle boundary disputes between squatters. Ten years earlier, there had been plenty of land for the taking; without too much trouble, Brodribb had taken 100 square miles. In those days, it was the squatting custom – recognised in the Act of 1836 – for neighbours to fix rough-and-ready boundaries by mutual consent. The Act of 1839, however, required the Commissioner to determine disputes about boundaries. This was a sign that a jostling for position was beginning. As the squatters began to look forward to improved terms of tenancy, the jostling grew rough. Thenceforward, Lambie had to make some difficult decisions. In June 1845, he upheld the rights of J. T. Moore against a man named Hibbard, who was nibbling at Burnima; but Hibbard's contention that there was one rule for the rich and another for the poor touched him on the raw.[1] About the same time, he was having trouble with William Bradley, who had bought the Cooma Creek run, but found when he tried to take possession of it that the seller's manager had transferred to neighbouring squatters two large chunks of land; the melancholy Robert Dawson had got possession of one chunk. Whatever the rights and wrongs in equity of this affair, Lambie decided that Bradley had no remedy in law. So it was the rich man, this time, who was left lamenting.[2] Naturally, the confrontations were not always so clear cut; Lambie from time to time had difficult decisions to make between the claims of contenders, neither of whom was rich, each of whom was deserving.[3] His letters reveal him as a man out of love with his duty as sole adjudicator, but resolute and honest in the performance of it.

In 1847, he and his fellow Commissioners found themselves working in a radically reconstituted environment. The Orders-in-Council of 9 March 1847, issued under an Imperial Act of the previous year, sealed the victory of the large squatters in their long and bitter battle with Governor Sir George Gipps.[4] The Orders-in-Council carried the squatters a good deal more than halfway along the road to ownership of their stations: thenceforward, their leases were to run for 14 years, with pre-emptive purchasing right over a square mile of land on each station: when the lease ran out, or whenever the lessee chose to sell out, he was to receive financial compensation for the improvements he had made.[5] From this

[1] Lambie/C.S.O.: 29 June 1845.
[2] Lambie/C.S.O.: 21 August 1845, with enclosures. Like J. T. Moore, William Bradley had stated his grievance in a petition to the Governor.
[3] Lambie/C.S.O.: 9 June and 9 August 1847.
[4] 9 and 10 Vict. c.104, 'An Act for Regulating the Sale of Waste Lands belonging to the Crown in the Australian Colonies'. The Orders-in-Council under this Act took effect by proclamation on 1 May 1847, and were amplified by regulations of 7 October. Only the provisions relating to the 'unsettled districts', such as Monaro, are referred to above.
[5] Most squatters received not a lease but a 'promise of lease', which carried with it all the privileges listed above.

time onwards, squatterdom felt secure enough to start civilising its way of life. Nobody remembered the disreputable origins of the word 'squatter'. Squatters were superior people.

The claims of the poorer licence holders, which throughout the previous decade had been a constant concern of Lambie's, received no recognition under the Orders-in-Council. On what terms the craftsmen, innkeepers, store-keepers, small graziers, and such-like little people managed to find places for themselves in the new order of society, would make a rewarding inquiry.[1] Our concern, however, is with the men who possessed, at the minimum, 4,000 sheep or their equivalent in cattle: nobody who possessed less was permitted to make application for a run. On the other hand, the age of unrestricted sprawl was closing: under the Orders-in-Council, 32,000 acres became the maximum size of a run. This limitation, however, was more apparent than real, because there was no impediment to one man possessing many runs. Not that he could possess them on such easy terms as in the past; the basic rental, for a run carrying 4,000 sheep or their equivalent in cattle, was £10, with an extra £2 10s for each additional 1,000 sheep or their equivalent.

The examination of applications for runs took time; but in September 1848 the *Government Gazette* published lists of the successful applicants for runs in each of the pastoral districts. Within two years, supplementary lists were published. On the basis of this information, the Monaro runs at mid-century are enumerated at the end of this chapter. Our total, 130 runs, is smaller than the gazetted total of 172, because we have excluded the still surviving extensions of the pastoral district – chiefly eastwards – beyond the three counties of the tableland. Moreover, our notes on the runs omit the rather questionable estimates of acreage and numbers of stock which the *Gazette* published. On the other hand, they indicate the nature of each pastoral enterprise (cattle, sheep, or mixed) and classify the proprietors as resident or absentee. Here could be the starting point of some useful research in the field of economic and social history.[2]

For present purposes, the enumeration of the runs serves a different purpose: namely, to put every pastoral property of the mid-nineteenth century into its proper place on the map. Easier said than done: in 1850, no map of the Monaro Counties existed.[3] This being so, it has seemed sensible to use the county maps of the present day; they are printed on the

[1] They probably continued to find grazing for their stock under the system of 'halves' or 'thirds', which had been widely prevalent throughout the previous decade. It would be worth while to inquire whether the acquisition of land by leasing it from large proprietors took root in Monaro, as it did on the Limestone Plains.

[2] The author is indebted to Miss Ruth Teale for making a search for the address of every holder of a run gazetted in 1848–50 in Monaro.

[3] The earliest map of the area so far discovered was drawn in 1854 by Surveyor Drake, on the scale of 2 miles to an inch. It shows the location of only about 20 runs.

scale of 2 miles to an inch and show the natural features in reasonable detail. Within the framework thus provided, the 130 runs are shown, in accordance with the description which their possessors gave of them 133 years ago (map 7, p. 47). These descriptions are, at the best, rough and ready: for one thing, the squatters had a financial motive – the larger the carrying capacity, the larger the rent – for understating the area of their runs: for another thing, they seldom, if ever, had a precise notion of the area and its boundaries. In defining the boundaries, they frequently referred to ranges or crests that are difficult if not impossible to find on the map, or to impermanent features such as cairns or blazes on trees, or – most frequently – to neighbouring properties no better defined than their own. Here, for example, is the description of Gejizrick, the run of Richard Brooks:

> Bounded in the east by Abram Brierly's run called Arable; on the west by T. V. Blomfield's run called Coolamatong, on the north by Wallace and Ryrie runs called Coolringdon, passing along the Spring Creek range; on the south by T. V. Blomfield's aforesaid, and Adam Brierly's run called Woolway.

To draw the boundaries of Gejizrick (now spelt Gegedzerick) as Richard Brooks conceived them to be in 1848, involves a good deal of guess work. The same is true, more or less, of all the runs. Notwithstanding these unavoidable shortcomings, the map embodies painstaking research and faithfully records the approximate size and location of each of the 130 runs.[1]

The shaded area of map 7 claims special attention. An earlier draft contained two such areas; but one of them, which showed the runs held in Monaro by Benjamin Boyd, was found to be expendable. Boyd was a financial wizard who sailed in 1842 from London to Sydney in his own yacht, *The Wanderer*. He held possession of liquid capital to the extent, at the very least, of £200,000. He arrived at a time when cattle and sheep and the pastures they were grazing on could be bought cheap; in Monaro, he bought sheep 'at 10*d*. per head with the stations given in'. That is the evidence of S. M. Mowle, who in 1845 enjoyed a sumptuous repast, with unlimited champagne, on Boyd's yacht in Twofold Bay.[2] An official report by Commissioner Lambie is substantially to the same effect: it shows that in mid-1844 Boyd was in possession of 14 Monaro runs totalling approximately a quarter of a million acres; for these 14 runs he was paying only four licence fees – a total of £40 per annum. By a rough calculation, that amounts to a rental of 1*d* for every 1,000 acres. Sardonic officials in Sydney made some similar calculations, and went on to say that Mr Boyd considered his runs to be his freehold.[3] Yet he would have done no good

[1] Mr Dan Coward deserves all the credit for this most interesting and useful map.
[2] S. M. Mowle, *Journal in Retrospect* (typescript copy, N.L.A.), p. 19.
[3] Lambie/C.S.O.: 3 May 1844 and enclosures.

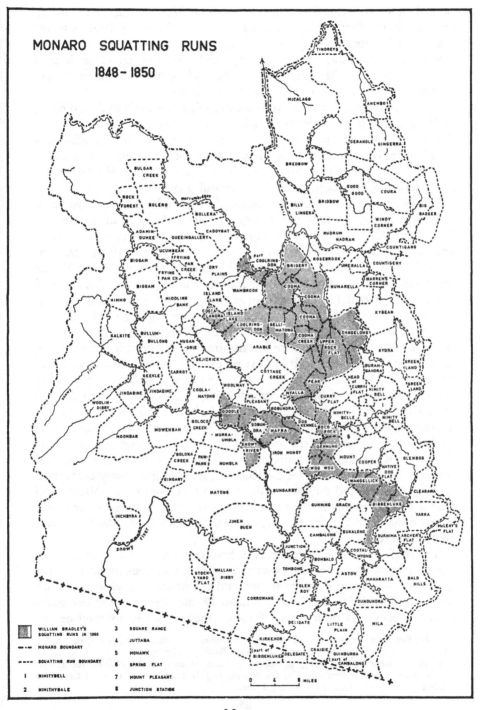

MONARO SQUATTING RUNS
1848 – 1850

Map 7

with them even if he had got them for nothing. He was a clever advertiser and money juggler; but he knew nothing about the use of land or about the men who used it. Very soon he ran into trouble. He cut wages and imported cheap labour – at least, he expected it to be cheap – from islands in the Pacific; but after a few years of false glory he went bankrupt. He merits no place on this or any other map of Monaro.

The shaded area retained on the map in its final draft shows the holdings of William Bradley at their peak, which was achieved in the decade that followed the Orders-in-Council, although even in the 1840s Bradley was running a close second to Boyd. He was a man of a different stamp, a man outstanding among his contemporaries as a *user* of land. His father, an army sergeant, had been the first successful grower of tobacco in New South Wales and had received substantial grants of land near Goulburn. The son retained that base and strengthened it; at the same time, he established an equally strong base in Monaro. He was a man of gigantic bulk and immense physical energy, who could shear a sheep or drive a team of bullocks as competently as any man in his employment. Although he was no less troubled than Boyd was by the labour shortage of the 1840s, he had a different way with his labour force; according to one of his managers, his rule was 'to pay and feed his men well and expect good results'. In the choice of his managers – men like W. A. Brodribb and James Litchfield – he showed not only intelligence but inspiration. We need feel no surprise at learning that his net profit in 1843 on every thousand sheep was £100; twice the rate for the colony, achieved when Benjamin Boyd was rushing towards bankruptcy. Bradley was, withall, a generous man, not only towards the men who served him faithfully and competently, but in support of any sensible project for the public good, whether it was railway building or improving the position of female immigrants. Caroline Chisholm, when she was expanding her energies on the latter cause, paid grateful tribute to him in her evidence before a parliamentary committee in London. 'He told me,' she said, 'that he approved of my Views, and that if I required anything in carrying my Country Plan into operation I might draw upon him for Money, Provisions, Horses, or indeed anything that I required.'[1]

The shaded area does not, however, hold its place on the squatting map merely in pious memory of a powerful and magnanimous pioneer. As the geological map (map 8, p. 49) shows, and as one can see when one looks down from the air on the chain of William Bradley's runs – or as one walks over the paddocks of any one of them – they are all established on the

[1] On Bradley see especially W. A. Brodribb, *Recollections of an Australian Squatter*, pp. 55–81; C. MacAlister, *Old Pioneering Days in the Sunny South* (Goulburn, 1907), p. 6; W. Davis Wright, *Canberra* (Sydney, 1923), pp. 49–51, 92; and the entry in the *A.D.B.*, vol. 1. Mrs Chisholm's testimony was given in London on 12 July 1847 to the House of Commons Select Committee on Colonization from Ireland.

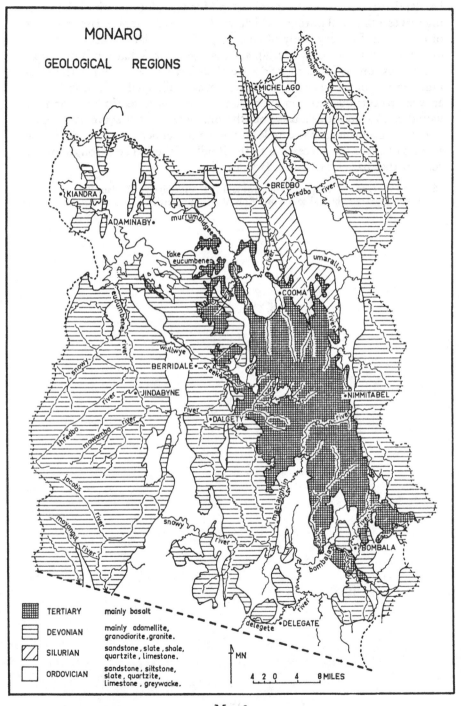

Map 8

basalt; that is to say, on the best land in Monaro. Not only that: the shading serves a practical purpose which will become apparent in later chapters of this book. In every study of land use, it is essential to get a clear view of the patterns of land ownership and management. What kinds of people got possession of Monaro land – we shall have to ask – when the mid-nineteenth-century squatting monopoly was challenged? To look for the answers to this question on every run that was gazetted in 1848 or 1850 would be too laborious a task; but the Bradley runs will serve as a representative and manageable sample. It will be our endeavour to illustrate on a series of maps the succession to the Bradley inheritance throughout the century that has elapsed since Bradley died.[1]

[1] See maps 9, 11, 12 and 14 (pp. 93, 104, 105 and 154, respectively). William Bradley died in 1868, two years after he had sold most of his Monaro properties.

Squatting

The list is arranged in five columns, as follows:

Column 1. The lease number, as given in 1848 and 1850, in the gazetted list of lessees and runs in the Maneroo District.

Column 2. Name of the run. (Here and in the text of the following chapters contemporary spellings are used.)

Column 3. Lessee or lessees.

Column 4. Residence:

      R = Resident in Monaro (total, 27)

      NR= Non-resident in Monaro (total, 71)

      U = Residence unknown (total, 32)

Column 5. Livestock:

      C   = Cattle (total cattle runs, 57)

      S   = Sheep (total sheep runs, 28)

      CS = Cattle and Sheep (total mixed runs, 44)

      H   = Horses (horse run, 1)

These labels are rather conjectural. The applicant for a lease was required to estimate the grazing capabilities of his run, and it has been assumed that he did so with reference to the animals actually grazing on his run when he applied for the lease. If this is so, we can deduce from the *Gazettes* a picture of the grazing use of land in Monaro in the late 1840s.

No.	Name of run	Lessee	Residence	Livestock
48	Adamindumee	Cosgrove and York	NR	C
34	Anembo	Joseph Bull	NR	CS
30	Arable	Abram Brierly	NR	CS
66	Archer's Flat	William Hibburd	R	C
133	Aston	Thacker and Co.	NR	S
109	Bald Hill Station	Pickering and Snape	NR	C
16	Bibbenluke	Ben. Boyd	NR	C
68	Big Badger	Edward Haslingden	R	C
82	Biggam	Wm. F. Kennedy	U	CS
132	Biggam	Wm. Thompson	U	CS
47	Billy Linpera	Cosgrove and York	NR	S
123	Bobundra	James R. Styles	NR	C
138	Bobundra	Charles Wright	NR	CS
19	Boco Rock	Ben. Boyd	NR	C
91	Bolero	Francis Mowatt	NR	S
5	Bollera	Wm. Barrett	NR	C
80	Boloco Creek	James Keirle	U	C
95	Boloka Creek	John McGuigan	U	C
41	Bambalo	Ronald Campbell	NR	CS
63	Bredbow	Patrick Harnett (estate of)	NR	CS
24	Bridbow	Stephen Burcher	R	C
54	Bulgar Creek	Thomas Eules	U	C
128	Bullumbullong	Joseph G. Thomson	U	C
4	Bukalong	John Boucher	R	S
108	Bungarby	Joseph Peters	NR	CS
89	Burnima	Thomas M. Moore	NR	CS
12	Burrangandra	James Binnie	NR	CS
126	Caddygat	James Stanley	U	C
20	Cambalong	Ben. Boyd	NR	C
46	Carrott	David Cassels	NR	S
8	Coolamatong	Thomas V. Bloomfield	R	S
147	Coolringdon	Wallace and Ryrie	R	CS

New arrivals

No.	Name of run	Lessee	Residence	Livestock
26	Cooma	Wm. Bradley	NR	CS
77	Cooma	James Kirwan	R	CS
164	Cooma	John Lambie	NR	H
131	Cooma Creek	Wm. Bradley	NR	CS
125	Corrowang	Stanton and O'Hare	U	C
118	Cootalandra	Stewart Ryrie, Jun.	R	CS
134	Cootalmyong	Thacker and Co.	NR	S
106	Cottage Creek	John Pendergrass	NR	CS
57	Countigang	Jeremiah Flynn	U	CS
78	Countigeny	James Kirwan	R	S
81	Coura	Thomas Kelly	R	CS
83	Craigie	Charles Lawson	R	C
155	Creewah	Conlon and Ryan	NR	CS
111	Curry Flat	Thomas Roberts	NR	CS
25	Dangelong	Wm. Bradley	NR	S
122	Delegate	George Simpson	R	S
37	Deligate	Robert Campbell (estate of)	NR	CS
3	Dog Kennel	Beard and Rolfe	NR	C
53	Doodle	John Eccleston	NR	C
103	Dundundra	John Nicholson, Jnr.	NR	CS
159	Dry Plains	Wm. Grahame	R	CS
158	Fryingpan Creek	John Fraser	U	C
160	Fryingpan Creek	Wm. Grahame	R	CS
59	Geekle	John Gore	NR	C
6	Gejizrick	Richard Brooks	R	CS
52	Gellimatong	Robert Dawson Jun.	NR	C
21	Gennong	Ben. Boyd	NR	S
156	Gerangle	John Cutmore	U	CS
127	Gingary	James Sherlock	U	C
152	Gingerra	Wm. Cooper	U	C
71	Glenbog	John Hosking	NR	C
117	Glenroy	Donald Ross	R	S
50	Good Good	Simpson Davison	U	S
137	Greenland	Gilbert Warren	U	C
162	Gunning Grach	Hughes and McIntyre	U	CS
139	Head of Curry Flat	John Williams	NR	C
112	Hugandree	Thos. L. and C. G. Robinson	U	C
172	Inchbyra	Ellen Woodhouse	NR	C
2	Iron Mungie	Beard and Rolfe	NR	C
45	Island Lake	David Cassels	NR	S
149	Island Lake	Wallace and Ryrie	R	S
38	Jimen Buen	Amos Crisp	NR	C
7	Jindabine	Richard Brooks	R	C
119	Jindabine	Stewart Ryrie, Jun.	R	CS
86	Junction	John Langhorn	U	C
98	Junction Station	James Marsden	NR	S
76	Juttabah	Thomas Jones (estate of)	R	S
168	Kalkite	Wm. Neale	U	C
115	Kybean	Charles Throsby	NR	C
33	Kydra	George Brown	NR	C
102	Little Plain	John Nicholson, Jun.	NR	CS
14	Mafra	Ben. Boyd	NR	S
114	Maharatta	Charles Throsby	NR	CS

Squatting

No.	Name of run	Lessee	Residence	Livestock
22	Matong	Ben. Boyd	NR	S
9	McLean's Flat	Henry Badgery	NR	CS
113	Micalago	Francis N. Rossi	NR	S
169	Middlingbank	John Neale	NR	C
94	Mila	Donald McPhee	U	C
35	Mohawk	Wm. Bowman	NR	CS
107	Moonbar	John Pendergrass	NR	C
36	Mt Cooper	Robert Campbell (estate of)	NR	S
58	Mt Pleasant	George Garnock	U	CS
116	Mt Pleasant	Charles Rootsey	R	C
70	Mowenbah	Wm. Holland	R	CS
15	Moyallon Downs	Ben. Boyd	NR	C
31	Murraumbla	Abram Brierly	NR	C
277	Myalla	Wm. Bradley	NR	S
120	Native Dog Flat	D. Rankin	R	S
124	Nimithybale	John Stanton	U	S
99	Nimitybelle	Donald McDonald	U	C
104	Nimitybell	Hugh O'Hara	U	C
121	Nimity Bell	Wm. Scott	NR	C
171	Nimity Bell	Bruce Reid	NR	CS
167	Nimmo	Austin Maley	U	C
39	Nudrum Nadran	Patrick Clifford	R	CS
88	Numarella	Joshua J. Moore	NR	S
161	Numbla	James O. Gorman	U	C
87	Pawpang	Daniel Lunn	U	C
90	Peak Station	Patrick Malady	R	C
44	Queeingallery	John Cosgrove	NR	S
49	Rock Forest	Peter Curtis	R	C
64	Rosebrook	Lawrence Harnett	NR	CS
11	Spring Flat	David Bell	NR	C
135	Square Range	Morgan Thornton	U	C
96	Stockyard Flat	Patrick McGuigan	NR	C
1	Tindreys	Henry C. Antill	NR	CS
136	Tom Bong	Wm. Whittakers	R	CS
51	Ucumbean	Daniel Driscoll	NR	C
85	Umeralla	Daniel Lucy	U	CS
28	Upper Rock Flat (includes 29 Lower Rock Flat)	Wm. Bradley	NR	S
60	Wallandibby	Alexander Gow	U	CS
55	Wambrook	Frederick Burchard	U	CS
140	Warren's Corner	John Waite, Jun.	NR	S
62	Windy Corner	Wm. Goodwin	NR	C
18	Wog Wog	Ben. Boyd	NR	C
93	Woolindibby	James McEvoy	U	C
32	Woolway	Abram Brierly	NR	CS
154	Yarra	Conlan and Ryan	NR	CS

2. Spoiling

From the time of Cook's first voyage until well into the nineteenth century, scientific discovery was a main objective of British policy in the Pacific Ocean and Australia. Enlightenment at the Royal Society and Admiralty, reinforced by Sir Joseph Banks' flair, gave the discoverers a wonderful start. Of the *Endeavour*'s complement of 94, one man in every ten was in the direct service of Banks. He brought back from the voyage an astonishing scientific harvest and the determination to promote further harvesting. That determination he was in a position to make good, for he was the fortunate possessor of health, wealth and social influence; he had still half a century of active life ahead of him and was to hold the president's chair at the Royal Society uninterruptedly from 1778 to his death in 1820. Not until the 1830s did his influence begin to fade.

In the age of Sir Joseph Banks and its afterglow men like Robert Brown, the botanist, John Lewin, the entomologist, and John Gould, the zoologist and ornithologist, won honour in the history both of Australian discovery and of European science. Moreover, in that happy age, the chasm between science and art had not yet opened up: sound draftsmanship was still a firm bridge between those two activities of the human spirit; Brown had working with him a superb botanical artist, Ferdinand Bauer; Gould had Elizabeth, his gifted wife; Lewin sought nobody's aid in illustrating his work. Another important bridge still held firm; Banks maintained a correspondence with every Governor of New South Wales from Phillip to Macquarie, and Alexander McLeay was secretary to the Linnaean Society in London before he took office as Colonial Secretary in Sydney. Practical men and observers, administrators, artists and men of science were still keeping company with each other.

All this was a bonus; but a period of comparative leanness was bound to ensue, until such time as the Australian colonies achieved the will and the means to make a substantial contribution of their own to art and science. The coming of the lean years proved to be roughly coincidental with the surge of the squatters into Monaro; yet even so, two scientific observers of some note, John Lhotsky and P. E. de Strzelecki, followed hard on the heels of the squatters. Both these men were originally Polish; but temperamentally and in their careers no two men could have been more unlike each other. This contrast adds piquancy to some historical debates of recent years: for example, was it Strzelecki, or was it Lhotsky, who made the first ascent of Australia's highest mountain?[1]

In the early 1830s, the King of Bavaria made Lhotsky a grant for zoo-logical and botanical research in South America and Australia. His arrival in Sydney in 1832 might have appeared the answer to prayer, for his scientific qualifications were unimpeachable and there was money on the estimates to fill the vacant position of Colonial Zoologist. Lhotsky badly wanted that position; but the government would not have him at any price. One wonders what went wrong. Lhotsky, perhaps, had uncovered a flaw, if nothing worse, in the system of public finance; this hypothesis could be made the starting point of a useful inquiry into the ends and means of financial policy at that time.[2] On the other hand, nothing may have happened beyond a clash of temperaments between Alexander McLeay, that stern, unbending Tory of Scottish breed, and John Lhotsky, that voluble, volatile, radical Pole. If this be so, what we want is not an historical inquiry, but an 'imaginary conversation'.

Thrown on to his own resources, Lhotsky wrote articles for the news-papers, sold firewood, vegetables and his own sketches, and ran into debt. By such means he was able not only to keep himself alive in Sydney but also to finance a three months' expedition to the Snowy Mountains. He set out on 10 January 1834 with four convict servants and a horse and cart. Into that cart he packed an astonishing cargo: things necessary for the 'health and comfort' of his party in the bush, such as powder and shot, tinder boxes and phosphorus, sugar and tea, sulphuric acid for making lemonade, plaster and physics, soap and candles, needles and thread, ropes and twine: things necessary for his scientific observations and collections, such as a telescope, hammers and chisels for breaking and shaping min-erals, nets for the capture of insects, forceps and 10,000 insect needles, anatomical instruments, a great quantity of paper for drying plants, drawing paper and colours, and a small reference library. So large and various a store of articles would have been too much for Lhotsky to manage, had not he acted on the theory that a man was free if he behaved like a free man. Under this treatment, his assigned servants became his

1 It has often been argued that Strzelecki gave the name Kosciusko to the slightly lower moun-tain which we know as Mt Townsend; but not until 1969 did anybody argue in a scholarly journal that the mountain which Lhotsky climbed and named Mt William the Fourth was the one which we call Kosciusko. See D. N. Jeans and W. R. Gilfillan, 'Light on the Summit: Mt William the Fourth or Kosciusko?', in *JRAHS* LV pt. 1, 1–81. N. A. Wakefield has come to a different conclusion on the basis of the same evidence as these two authors used: one awaits the publication of his rejoinder.

2 It stuck in Lhotsky's throat that the government was raising more money than it was spending in the colony – and this at a time when the frontier districts were anarchical, in the literal sense of that word; when the assigned servants were inadequately clothed and too often unprovided with medical care; when the position of Zoologist was kept vacant; when there was no postal service even to the Limestone Plains. His contention that public parsimony was at odds with the needs of the developing colony deserves to be taken seriously.

fellow-workers: Walker was the hunter, bird stuffer, shoe and harness maker; William was the store-keeper; Relf was the groom, tailor and plant collector; young Paddy was the insect collector. After a hard day's work, they all played cards together, or Lhotsky read *Robinson Crusoe* to them.

On his return to Sydney, Lhotsky wrote and got quickly into print an account of the expedition. His book is a delicious travel story.[1] The style is simple and lively; when Lhotsky quotes a word or two of Latin or Greek, he does not do it for show, but to give point to his meaning; when he coins a new English word, he entrances us – on a cool morning, he finds the air 'alpester'; on a humid morning, he finds it 'mizzly'.[2] The style reveals the man: Lhotsky's emotions of anger, compassion, love and joy ripple the surface of his thought and prose. He is at times inconsequential; he hates the squatting monopoly of land, but he admires individual squatters and foresees a place for counts and dukes in Monaro of the future. At the mineral spring near Cooma, he achieves inconsequence on the heroic scale; this place, he tells himself, will become a new Bath, where kings or presidents of the Antipodean lands will build their palaces and take the waters. Alas, it did not so happen. A few years later, shepherds and stockmen of the neighbourhood found that the mineral water went well with rum. One night, they poured in a whole keg of rum. As if resenting such outrage, the spring spouted mud.[3]

Lhotsky's absurdities are endearing and do not in any way diminish the historical value of his book. This, as will appear later, is considerable, for nowhere else can we find so vivid a picture of poor people's Monaro.[4] On the other hand, the book was not, nor was it intended to be, a work of science. One may suppose that Lhotsky had it in mind to work at his collections and embody the results in scientific papers, if not in a book; but this he was never in a position to do, because of his poverty. He sold parts of his collection to pay his debts; he gave some parts away; perhaps he had to leave some parts lying in a hovel or shed, when he gave up the struggle to find an Australian foothold and in 1838 sailed for London. Surely he had deserved kinder treatment? A genus of Australian plants (*Lhotskia*) and one of fishes (*Lhotskya*) testify that he was truly a discoverer, and scientists of our time, when they can be persuaded to read his travel story, consider his observations, given the state of knowledge in his

---

[1] *A Journey from Sydney to the Australian Alps . . .* , by Dr John Lhotsky (Sydney, 1835).

[2] *Ibid.* pp. 71, 91. For another superbly apt word of his own coining see 'hudibrasising' on p. 73; for an equally apt Latin quotation see *Meminisse juvabit* in the footnote to p. 76.

[3] H. W. Haygarth, *Recollections of Bush Life in Australia* (2nd edn., London, 1861), p. 29. Mineral water from Lhlotsky's spring is today being sold commercially.

[4] Word had gone round among rich people that Lhotsky was a man to shun. He mentions three rich people only – Robert Campbell, William Bradley, and Dr Reid – who gave him help and hospitality, in person or by deputy.

Spoiling

time, to be accurate and intelligent.[1] What most impresses the layman is his modesty. He keeps telling us that he has not skill enough to identify this pebble, or time enough to make up his mind about that plant; he confesses he can give only a few hints to Australian hydrographers of the future; he says that he does not know why his mineral spring is lethal to insects; he puts forward an hypothesis to explain the great chain of treeless plains, and then points out its inadequacy. His highest hope, he says, is 'to lay some slight shade of colour upon a part of the undefined blank of the map of Australia'.[2]

Strzelecki, who came to Australia two years after Lhotsky left it, was a man of larger ambition. After a false start in his own country he achieved recognition both in the world of science and in English society. The controversies that have arisen about his character and career do not concern us;[3] but let us meet the man in a passage of his prose. Whether or not we approve the style – some people would call it magisterial, others would call it pretentious – we cannot but agree that he states his argument with power. In any anthology of the literature of land use and conservation, his statement would merit a prominent place.

He is submitting a report to Governor Gipps on his recent travels in the Australian Alps and Gippsland.[4] In discussing 'the extraordinary droughts' of this area, he says that they are in part the product of atmospheric phenomena that occur in the similar latitudes of other continents; but he then goes on to say that these similarities do not tell the whole story.

> The drought, however, here in New South Wales seems to me to have an additional cause to that or those which elsewhere occasion extraordinary dryness of the soil: namely, the alteration which colonisation impresses on its surface; the herbaceous, high and thick plants; the continued forest; the underwood; the brush, which so well clothed the crust and sheltered the moisture, have disappeared under the innumerable flocks and axes which the settlers have introduced. The soil, thus bared, was and is, as it were, abandoned by a most prejudicial practice, to the constant and periodic wilful incendiarism, which, instead of producing the expected and former herbage and vigour of the soil, in fact only calcines its surface and eradicates even the principle of reproduction . . .
>
> The analogy, the admirable harmony and correspondence which, in the economy of the globe, is everywhere established between the ambient air, the

[1] Communications by Dr Germaine Joplin on the geological observations recorded in his travel book, and by Dr Maisie Carr on the botanical observations.
[2] Lhotsky, *A Journey from Sydney* . . ., pp. 54, 55, 96, 97, 113.
[3] On the Strzelecki controversies, see H. M. E. Heney, *In a Dark Glass* (Sydney, 1861), a most interesting book, which received a hostile review from G. Zubryzcki in *Hist. Stud.* II, no. 41 (1963), 135–7.
[4] Enclosed in a despatch from Gipps to the Secretary of State, 28 September 1840, and printed in H. of L. Sess. Papers, 1841(85), pp. 12–19.

superficial state of the soil, and the constitution of organic and inorganic bodies placed upon it, finds itself altered here. The disturbance of this analogy extends the dryness of the soil, augments its aridity, and favours the percolation of waters, or their evaporation. Human industry hitherto but increases the evil; the rotation of crops, dams to arrest the torrents, reservoirs to contain and preserve them, artesian wells to bring to the surface these innumerable hidden sources, irrigation, manuring, artificial grasses, have yet to make their appearance in this colony. The increase of grazing flocks, which in many cases overstock the pasture, inferior agriculture, seeds more inferior still, are alone progressing from day to day, enervating the surface and assimilating its best parts to the rest already sterile.

The idea that soil cover is precious and its destruction dangerous was not, of course, a new discovery by Strzelecki; his elder contemporary Alexander von Humboldt had made it a commonplace among students of the earth sciences.[1] Long before von Humboldt, many thoughtful people had shown particular concern for the protection of forests. In seventeenth-century France, that concern had found expression in Colbert's Forest Ordinance, to which we may trace the origins of a superb institution of contemporary France, the department of *Eaux et Forêts*.[2] Two thousand years before Colbert, the same concern had found expression in Plato's strange dialogue, *Critias*. Why, Plato asks, are the Athenians unable nowadays to keep all their fighting men in the field, as they had been able to do in time long ago, when they were at war with Atlantis? In answering this question, Plato paints a vivid picture of the ruin that the Athenians have inflicted on their soil, and on themselves, by destroying their mountain forests. The Attica of our ancestors, he says, was the richest land in all the world; but our Attica is no more than 'the bones of a wasted body'.

That such ideas have been in the minds of thoughtful people for a very long time does not, of course, prove them to be true; in some of their formulations, they may be erroneous. For example, in Strzelecki's time many people used to say that forests bring rain; but in our time few, if any, scientists would repeat that assertion.[3] That forests help to conserve water, after the soil has received it as rain or snow, is a different proposition; but it cannot be profitably discussed unless careful consideration is given to the almost infinitely variable conditions of different places and times. Even if thus put into context, the proposition seldom gives so good a start to the discussion as a simple question would give: what,

[1] Strzelecki's book, *Physical Description of New South Wales and Van Diemen's Land* (London, 1845) exemplifies von Humboldt's spacious framework of thought.
[2] *Eaux et Forêts* has some significance in Australian history because C. E. Lane Pool, an eminent and influential Australian forester, received part of his training there.
[3] It seems now to be accepted by the great majority of hydrologists, meteorologists and other specialists that man's activities on the surface of the earth have no effect at all on the macroclimate, but may have some effect on the micro-climate. Strzelecki, however, asserts the opposite (see *Physical Description*, p. 239).

Spoiling

in these or those conditions, is the most efficient soil cover and water conserver? In Attica, as in most other Mediterranean regions, no adequate substitute was ever found for the forest communities that had been plundered; but on English downlands grass makes very good cover. Nor can one envisage better cover on the Monaro tableland than a thick sward of clover and phalaris.[1] At the snowgum level in Monaro the situation is different; there, systematic observations of vegetational change on fixed reference areas were initiated some years ago and are still proceeding. Above the snowgum level, where alpine herbfields and sphagnum bogs are natural water conservers, the situation is different again, and has been studied throughout recent years with the same purposeful and patient attention to detail.

It goes without saying that Strzelecki could not possibly have envisaged, more than a century ago, the precise objectives and techniques of ecological research in the mid-twentieth century. Australians have good reason to ponder his forceful denunciation of ruinous land use, his proposals for reform, and the criteria of sound practice which he states or implies: the fertility of the soil, he says in effect, should be harvested, not mined. We, however, are pursuing an inquiry into land use in Monaro, and can find little information in Strzelecki's sweeping generalisations. For example, he tells us nothing at all about the succession of wetter and dryer periods; on the contrary, he took it for granted that the appalling drought, during which he made his journey from Yass to Corner Inlet, was the normal state of the weather in that part of the country. In his denunciation of incendiarism, he took it for granted that the incendiaries were white men; whereas Hovell before him and Townsend after him took it for granted that they were black men.[2] As we have seen, the Aborigines had been in the fire-raising business for a good many thousands of years. Strzelecki, when he set out on his journey, had been in Australia for less than one year. This was time enough for a man of his wide reading and experience to frame some useful questions about land use in New South Wales, but not time enough for him to make a precise report upon the conditions of any particular region. Monaro, besides, lay to the east of his line of march. The time that he spent in Monaro amounted to no more than the time he spent on the alpine summit which he climbed – at the most, an hour or two.

We need testimony based on local knowledge. Fortunately, a young Englishman, Henry W. Haygarth, has given us illuminating testimony.[3]

[1] One must not, of course, ignore the possibility of the new pastures proving more vulnerable to disease than the more complex native grass communities which they are replacing. See pp. 188–9, below.
[2] On Hovell see p. 73, below, on Townsend p. 6, 22, above.
[3] Haygarth's book, *Recollections of Bush Life in Australia during a Residence of Eight Years in the Interior*, was published in Murray's Home and Colonial Library in 1848 and re-issued in.

59

New arrivals

Haygarth spent eight years of drought, slump and recovery on a run in the Squattage District of Monaro. He entered with zest into the squatting life and at the same time observed and recorded its peculiarities – for example, its peculiarities of speech: stockhorse; three-cornered horse; gully-raking; splitting; cutting-out; overlanding; a creek that is running a banker; a man who has no flies on him – as Haygarth writes, we hear the talk of the cattle station. Incidentally, we see the word 'station' rapidly changing its meaning.[1]

Haygarth's evidence is nonetheless vivid because of the confusions contained in it; he manages to be at one and the same time in love with primaeval nature, with 'improvement', with 'sport'. In a passage much quoted by nature conservationists, he pictures a sportsman making the first journey ever into unspoilt country. Then he pictures the spoiling that follows.[2]

> The most spirit-stirring sight which the sportsman can witness is the first view of a new pastoral district; and to the lover of the picturesque perhaps this is the most beautiful scene that Australia can afford . . . Plains and 'open forest', untrodden by the foot of the white man, and, as far as the eye can reach, covered with grass so luxuriant that it brushes the horseman in his saddle; flocks of kangaroos quietly grazing, as yet untaught to fear the enemy that is invading their territory; the emu, playfully crossing and recrossing his route; the quail rising at every step; lagoons literally swarming with wild-fowl – these are scenes reserved for the eye of the enterprising settler, or the still more enterprising 'overlander'.
>
> Then mark the change that follows hard upon discovery. Intelligence of the new country reaches the settled districts, and countless flocks and herds are poured into the land of promise. It is divided into stations, and 'improve-ments' are everywhere erected on it; disputes arise, and a commissioner is appointed to settle them; bushrangers are 'out', and mounted police are sent

1861. The author gives no particulars about himself; but from published lists of Eton and Oxford, and an obituary in *The Times* (1 January 1903) the following chronological landmarks have been identified:

1810:	Birth
1820s:	At Eton
1839–46 (approx.)	In Monaro
1847–51:	At Exeter College, Oxford
1851:	Ordination in Church of England
1859–1902:	Vicar of Wimbledon (with other ecclesiastical offices)
1902:	Death

[1] In the 1820s, 'station' meant a night's camping place during a journey (see Hovell's journal, *JRAHS* VII (1921), 339). In the 1830s, landowners in the settled districts sent stock to 'out stations' beyond the limits of location. Occupants of out-stations made a distinction between the 'head station' and its 'sheep stations', or 'out-stations'. In the 1840s, the whole property is called (as sometimes by Haygarth) 'a station', although in 1848 its official designation is still 'a run'. By the 1850s, any substantial pastoral property, even if it is situated within 'the settled districts', is called a station.

[2] *Recollections of Bush Life*, pp. 120–1. The passage is quoted in a desk diary for 1969, entitled *The Early Australian Scene: Animals tell the Story*, published by the National Trust of Australia.

60

to hunt them down; the wild blacks, indignant at the cool occupation of their territory, spear the cattle, and the settlers retaliate. The governor establishes a 'protector of aborigines', who perhaps has most need of protection himself. To some the regions bring wealth, to others disappointment, while Anglo-Saxon energy at last triumphs over every obstacle. But Nature, as if offended, withdraws half her beauty from the land; the pasture gradually loses its freshness; some of the rivers and lakes run low, others become wholly dry. The wild animals, the former peaceful denizens of the soil, are no more to be found, and the explorer, who has gazed on the district in its first luxuriance, has seen it as it never can be seen again.

Haygarth was an experienced and sensitive eye-witness; but his training was literary and theological, not historical or scientific. So let us treat his testimony simply as the agenda for an inquiry into spoiling under the three heads that his sentences suggest: the spoiling of plant life, of animal life, of human life.

Up to and beyond the middle of the nineteenth century, the study of Australian plants continued to make great strides: Robert Brown, Alan Cunningham, Baron von Mueller – it is a splendid botanical trinity. These men saw Australia as a newly discovered treasury of plants; they saw themselves as explorers, enumerators and classifiers of the treasures. They were systematists, not ecologists.

We may save ourselves some confusion of thought if we spend a minute or two considering the word 'ecology'. The word is a newcomer to the dictionary.[1] Broadly, it signifies a 'natural' community of plants and animals inhabiting the same 'natural' home. It expresses a point of view: namely, that the lives of all the inhabitants – trees, shrubs, herbs, animals vertebrate and invertebrate, creeping things in the grass, bacteria in the soil, insects on the wing – are all closely intermeshed; indeed, so closely intermeshed, that no individual life can be understood except in relation to the lives of fellow-members of the community. Much can be said in favour of this point of view; but the word 'natural' raises some difficulties, particularly in relation to the activities of man. Man, undeniably, belongs to the world of nature; but he also ascribes to himself dominion over nature; we learn this both in the first chapter of the Book of Genesis, and in recorded human history. Ecologists seem quite often to be in two minds about how best to cope with this duality of human nature. Some of them announce their intention of putting man into the centre of the ecological

[1] In the *Oxford English Dictionary*, the first quoted use of the word (then spelt oecology) is by Haekel in 1877. Ecology, like economy, stems from *oîkos*, meaning home. Cf. Sir Arthur Tansley, *Introduction to Plant Ecology* (1962 ed.), p. 15: 'In its widest meaning, ecology is the study of plants and animals *in their natural homes*; or better perhaps, the study of *their household affairs*, which is actually a secondary meaning of the Greek word'.

picture, but in practice do the opposite.[1] Others envisage an 'ideal' situation, a 'climax' achieved by nature in the absence of man; but sooner or later they bring man in as an intruder, as the maker of 'disclimaxes'.[2]

Australian ecologists, in the main, have kept themselves aloof from these puzzling theoretical issues and have given themselves whole-heartedly to field research. In so doing, they have gathered a harvest of knowledge which is serviceable both to natural science and to history. In the present inquiry, we are in debt to them on two counts: first, for their comprehensive and precise enumeration of the plant communities as they were before the white men arrived; secondly, for their elucidation of the processes of change which the white men initiated.[3] Inevitably, they have less information to give us about the gathering momentum of change from one historical period to the next. From their own direct knowledge, they can tell us what man in Monaro has been doing to the pastures during the past three or four decades; but we cannot expect them to tell us what the squatters did to the pastures.

It follows that we have to look for earlier testimony. As we have seen, it is not to be found in the writings of the early systematists; so we must look for it later in the nineteenth century. Ideally, it should come from a witness who had had some acquaintance with the squatting life when he was a boy, and had steeped himself when he became a man in the teaching and method of Charles Darwin. By a rare stroke of fortune, the witness we need is forthcoming. His name is A. G. Hamilton. His testimony is recorded in a long paper 'On the effect which settlement in Australia has produced upon indigenous vegetation'.[4]

Hamilton pursues his inquiry under three main heads: first, what has been done to the indigenous vegetation by direct action of the men; secondly, what has been done to it by the animals which accompanied the men; thirdly, what has been done to it by the plants which accompanied them. Under each head, he systematically enumerates and examines the ramifications of the impact and the consequential change. For example, under the first head, he starts by examining the various motives and methods of deforestation; he then essays a quantitative estimate of its extent; he then discusses *seriatim* the consequential changes in the understory of vegetation, in the storage of water in the soil, in the surface flow of water, in the productivity of the soil: finally, and for good measure, he examines the consequences of the efforts made by the white men to prevent

---

[1] Sir Arthur Tansley, in the book cited above, appears to do this.
[2] See Fraser Darling's statement in *Man's Role in Changing the Face of the Earth*, ed. William L. Thomas (Chicago, 1956), pp. 407–8.
[3] I have already acknowledged my debt to A. B. Costin. Cf. the papers of L. D. Pryor and R. M. Moore in *Canberra – A Nation's Capital*, ed. H. L. White (Sydney, 1954). Moore has recently edited a useful symposium, *Australian Grasslands* (A.N.U. Press, 1970).
[4] *Journ. Roy. Soc. NSW.* XXVI (1892), 178–239.

or check bushfires.[1] Under the second and third heads of the paper, he is equally thorough in the discussion of salient issues. Through his eyes we can see the operation of a botanical Gresham's law, under which pastures that have been too heavily grazed go from good, to bad, to worse; we can see the hoof of an ox making a hole in the ground, and that hole starting a flow of water, and that flow of water carving a deep gulley through which immense quantities of soil are carried down hill; we can see nostalgic Scots planting thistles around their huts and the wind scattering the thistle-down far and wide. Even when he may seem to be reporting trivial occurrences, Hamilton is invariably concerned with their intermeshing in 'the network between the animal and vegetable kingdoms' and – one may add – between the elements of soil and climate.[2] His observations and reflections possess both local and global significance: to cite one example out of many, his sombre arithmetic of the net losses of fertility which the wool and meat industries inflict on the soil – when the owners of the grazing animals possess no science and no good habits of pasture management – would have been just as much to the point if he had been examining the use of land, not on the tablelands of New South Wales, but in the Scottish Highlands.

We shall be meeting Hamilton from time to time. His paper could be made the starting point of a dozen or more specialist monographs. However, our inquiry is non-specialist and we have to cover a great deal of ground quickly. It is time for the users of the land to come again into the story. In the early decades of colonisation, most of them were uninstructed not only in botany but also in the practical arts of tillage and pasture management. Moreover, agricultural experience in the British Isles, when by rare exception a rich man or a poor man possessed it, could prove irrelevant or positively misleading under Australian conditions. It was only by trial and error that the settlers around Sydney came painfully and precariously to terms with their new environment. In this process, they sorted themselves out – or were sorted out by stern authority – both sociologically and topographically: by and large, the poorer land users tilled alluvial soil in pockets along the coast and in the valleys of rivers and creeks, while the richer ones depastured cattle and sheep on the grass of the open forests and lightly timbered plains. The grass did not make the kind of sward that they had known in their homeland, but grew tall, frequently in clumps or tussocks. There was a lower story of short

[1] Hamilton says that the settlers increased their efforts to prevent fires after they had fences to protect and that, even in the days of shepherding, their fire precautions were in marked contrast with Aboriginal practice. At the same time, he is aware that they often burnt dry grass in order to get a green bite for their livestock.
[2] Hamilton, *On the Effect Which Settlement . . .*, p. 202; also pp. 192–200, where he carefully reports the controversy about whether or not forests bring rain, and concludes that they do not, thereby rejecting what was still, perhaps, the majority opinion. See p. 58, above.

grasses; but these, to begin with, attracted little if any attention.[1]

After a few decades of trial and error, the graziers, with few exceptions, still remained ignoramuses in pasture management. However, they believed that they knew their own business. Land and labour could be got cheap. Capital could be accumulated quickly provided one belonged to the inner ring of shippers and sealers, traders and financiers. Markets were chancy for the man who tilled the soil, but far less so for the man who depastured stock; when manufacturers in Yorkshire became eager bidders for Australian wool, the prospects of expansion began to seem limitless. Meanwhile, the natural increase of the flocks and herds continued from year to year. These two elements of demand and supply have usually seemed sufficient to explain the geographical sprawl of the pastoral industry; but equal weight must be given to a third element, namely, mismanagement of the pastures and their consequential impoverishment. In the early years of colonisation their fragility could hardly have been anticipated, because they had never before been continuously and heavily grazed. Marsupials, having no hooves, tread delicately; but sheep tread firmly and cattle stamp hard: marsupials graze lightly; but drought-afflicted sheep will eat the grass to its roots and then pull up the roots and lick up the seed. Under such rough usage, the better native species were succeded by inferior species and by weeds which had made the passage from Europe as stowaways. The graziers believed, with good reason more often than not, that the nastier stowaways and the nastier native plants – for example, wire grass and corkscrew grass – were thriving most vigorously. They also denounced a nasty caterpillar – probably it was the army worm. In good seasons they coped well enough with these afflictions; but when drought came they found their pastures overstocked. To sum up: it was not only the pull of opportunity, but also the push of necessity that turned them into a race of nomads, everlastingly in quest of 'better country farther out'. As users of the land, if not of the Bible, they were blood brothers to the trekking Boers of South Africa.[2]

By the late 1820s, they were infiltrating Monaro.[3] Ten years later, they were infiltrating the Port Philip District and intermingling there with other

[1] In this and the following paragraphs use is made both of the botanical papers cited earlier and of T. M. Perry, *Australia's First Frontier* (M.U.P. 1963). The tall grasses referred to were normally an association of *Themeda australis, Poa caespitosa* and *Stipa aristiglumis*. The graziers were especially in love with the first of these three, which they called Kangaroo grass. It is not always easy to find the correct botanical equivalent for the common name of a grass – for 'oat grass', for example, which had high prestige in New South Wales, or for 'silk grass', which was loathed in Tasmania. 'Wallaby grass' was one of the *danthonias*; it was palatable and nutritious.

[2] Cf. W. K. Hancock, 'Trek', in *Econ. Hist. Rev.* 2nd ser., x (1958), 331–9.

[3] In a letter to his solicitor dated 24 March 1853, Richard Brooks of Gegedzerick declared that he had held possession there since 1827. Cf. Arnold Harris, *Berridale: A History of the Parish* (1935), p. 2.

infiltrators, in flight from the impoverished Tasmanian pastures.[1] So the sprawl continued. And so, as Haygarth saw,[2] it was bound to continue, until the desert blocked it and better management of the better country made it superfluous. The beginnings of better management will come into the story a little later; but meanwhile there is nothing good to be said about what the squatters did to the pastures. In Monaro as elsewhere, they made one blade of grass grow where two had grown before.

What they did to animal life seemed to Haygarth plain extermination: enter 'the sportsman', exeunt the animals. But the sequence of events was not so simple as that. If Haygarth had returned to Monaro twenty years later, he would still have lamented the disappearance of the emus; but he would have seen grey kangaroos in such immensely augmented numbers that the settlers were organising hunting drives to keep their increase in check.[3] Similarly, if he had moved quietly through the bush at night, he would have observed far more rat kangaroos, native cats, koalas and other small marsupials than he had ever seen during his eight years as a squatter.[4] Another twenty years later, if he had made a second return visit, he would have noticed that the populations of large and small marsupials had fallen again; but he would have seen as many parrots as there had ever been and, possibly, a good many more galahs.

To explore and explain these swings of population is a task that an ecologist and an historian might profitably tackle in partnership, provided they had a few years to spare. The best that we can do is to put up signposts to one or two promising paths of research. So let us, to begin with, look a little more closely at Haygarth's 'sportsman'. A good many squatters kept hunting dogs and nearly all of them had guns. On 8 March 1841, Terence Aubrey Murray of Yarralumla went hunting in the Brindabella mountains with a few friends. They climbed to the summit of Pabral Peak and found the outlook so sublime that they agreed with each other to say a prayer.

> Before, however, we had time to carry this proposal into execution, the black who was with us gave the alarm from a thicket that there was game in view and we all started in pursuit. He had discovered a few lyrebirds which were a short distance from us. They were singing most beautifully, and we listened

---

[1] See Bride, *Letters from Victorian Pioneers* (pp. 33–5) for an exceptionally good account of the succession of pastures, by John G. Robinson.

[2] *Recollections of Bush Life*, p. 146, Haygarth says that overlanders perform the function of shifting stock from exhausted pastures to fresh pastures.

[3] See pp. 113–14, below.

[4] Hamilton, *On The Effect Which Settlement . . .*, p. 213, quotes evidence of their extraordinary increase in Victoria. In Monaro there were large numbers of them on stations in the 1870s (communication from Mr R. B. Rose). See p. 113–14 below.

with surprise and pleasure. I did not think that I ever heard birds that equal them in melody. They saw us, however, and escaped. We then descended the mountain, a difficult task. On the way down we saw more lyrebirds, at two of which I had two shots with my rifle.[1]

The spirit of the sportsman was just as much alive in the breast of that lonely, pious Scot, Farquhar McKenzie. On his journey into Monaro he relieved the tedium of driving tame sheep by shooting wild animals. For example:

> When out shooting duck at Bredbow river I saw a number of those curious animals the Platibus [*sic*] and fired at them repeatedly – but whether they dived at the flash of the pan or sank after being shot I know not.

Other people, including Lhotsky and Haygarth, noted the extraordinary speed of the platypus in 'ducking at the flash'.

Lhotsky was not a sportsman, but he had two good reasons for sending his man Walker out with the gun; first, he needed meat for his small party; secondly, he needed specimens for his zoological collection. The squatters had no zest for zoology, but were apt to favour a stuffed platypus as a sporting trophy on the hut wall, or as a present to a friend in the city. Besides, kangaroo tail soup added variety to a monotonous diet, while the tanned skin of a kangaroo made a good rug for the dirt floor or, in the freezing winter nights, for the bed. Thus there were all sorts of reasons, in addition to the sporting impulse, for killing the native animals. Nevertheless, we should beware of exaggerating the extent of the killing. In those days, the commercial traffic in the skins and flesh of marsupials, if it existed, was not yet big business. Besides, the spaces of Monaro were vast and the human population sparse. It seems reasonable to envisage the animal population, in so far as it was mobile, retreating uphill from the discomforts and dangers of life at the lower levels.

If this be true, our problem in the squatting country is not the initial decrease of the animal population, but the subsequent increase of some species. This problem may seem to belong to a later time; but two things which were happening in the squatting period have a bearing on it. First, as we have already seen, the cattle were eating out the tall grasses: the squatters lamented their passing, but shorter grasses proved more acceptable not only to the sheep but also to the grey kangaroos and wallabies. Secondly, both the larger and the smaller marsupials began soon to enjoy a new freedom from the attacks both of the Aborigines, whom the squatters pushed off the land, and of the dingos, against whom they waged implacable war.[2] In these and other ways that nobody had foreseen,

---

[1] Gwendoline Wilson, *Murray of Yarralumla* (Melbourne, O.U.P., 1961), p. 131, quoting Murray's diary. Pabral Peak is now called Mt Coree.

[2] On this, there is striking unanimity of testimony from many intelligent observers: see e.g. Edward M. Curr, *Recollections of Squatting in Victoria* (Melbourne, 1883), ch. 17, and the evidence quoted in W. A. Hamilton, *loc. cit.*

squatting prepared a habitat in which the native animals multiplied. Later in the century, fences, foxes, rabbits and people wrecked that habitat. The larger marsupials retreated again uphill and the smaller ones survived, at best, precariously.[1]

The 'Menero tribe', Lhotsky wrote in 1834, 'is already very weak, consisting of about fifty men; they are entirely tame (indeed not civilised but corrupted) . . .'[2] Events, it would seem, had taken a disastrous turn for the people who, eleven years earlier, had so timidly approached Captain Mark Currie.

It is difficult to measure the extent of their disaster. Lhotsky did no more than report what one man had told him about the population of one tribe. Official counting of the population of all the tribes of the Pastoral District did not begin until 1842, when Commissioner Lambie submitted to the Colonial Secretary the first of his annual censuses. In that year, his estimate of the total Aboriginal population was 798; seven years later, it was 600. These estimates, as he took pains to point out, were rough and ready; nevertheless he felt no doubt that they told a sad story.[3]

The story, however, will not make sense so long as its first chapter is missing. We must make an estimate, conjectural though it is bound in some measure to be, of the Aboriginal population in the late 1820s, when the first squatters rode in with their cattle and sheep. Thanks to the combination in field research of the ethnographical, demographical, ecological and historical approaches, we possess surprisingly firm knowledge of tribal populations in every geographical region of Australia. They varied upwards or downwards from an average of approximately 500. If we take 500 as the multiplier, and then look at our maps, our figuring should be reasonably realistic. Three-counties Monaro was in area roughly coincidental with the territory of a single tribe, the Ngarigo: consequently, the total Aboriginal population of this area cannot have been much above 500 – about one person to the square mile. We, however, are at present concerned, as Lambie was, with a much larger area, the Squattage or Pastoral District of Monaro. Since this area contained five tribal territories, our estimate of its total population, before the squatters invaded it, is approximately 2,500.

[1] A brief retrospective survey of zoological change in closely comparable country is given by F. N. Ratcliffe and J. H. Calaby in *Canberra – A Nation's Capital* (ed. H. L. White, Sydney, 1954), pp. 178–87. See pp. 113–18, below.

[2] Lhotsky, *A Journey from Sydney . . .* , p. 106.

[3] Lambie's censuses of the Aboriginal population were submitted in January of each year. They were enclosed with despatches from the Governor to the Secretary of State and can most conveniently be consulted in H.R.A. In consulting them, one needs to bear in mind the changes in the boundaries of the Pastoral District of Monaro, and the other good reasons for using them with caution.

New arrivals

If we look now at the earlier and the later chapters of our story, we can see that the curve of depopulation was flattening out during the 1840s. To be sure, there was a disproportionate and ominous decline in the numbers of women and children; but the decline in total numbers was not, for the time being, catastrophic. Catastrophe had already occurred. Between the late 1820s and the early 1840s, the Aboriginal population had fallen from a figure well above 2,000 to one well below 1,000.

So we must try to understand how it came about that the tribes were cut down to nearly one-third of their size during the first fifteen years of the white invasion. Physical conflict is inevitably prominent among the explanations that come to mind. In the neighbourhood of Port Jackson, conflict had become sporadic before twelve months had passed since the arrival of the First Fleet. Thereafter, as the pastoral frontier moved inland, similar sequences of aggression, resistance and retaliation repeated themselves with nauseating frequency; how nasty those sequences were we can see – to cite one source of evidence among many – in *Letters of Victorian Pioneers*. But in Monaro, no evidence at all of physical conflict can be found, either in the diaries and letters of pioneers, or in official records.[1] Monaro, we are bound to conclude, was an exception proving the rule. There, as everywhere else, the white men took the land, and some white men took the women; but there, as almost nowhere else, resistance and retaliation did not ensue. Quietly, the Aborigines submitted. Quickly, they became, to quote Currie's word once again, 'domesticated'. It is hard to find any reason for this breach of the normal sequence, except, possibly, in the time factor. In the neighbourhood of Sydney, it had taken the Aborigines years of suffering to learn the lesson of their own helplessness; but then – as W. E. H. Stanner has put it – 'something breaks in the fabric of native life, and from every side the aborigines, unforced, begin to flock into the settlement'.[2] Thereafter, as the white men swarmed into the interior, one tribe after another learned the same bitter lesson in the same hard way. But not the tribes of Monaro: they, we must assume, learnt their lesson from the experience of others and made their submission, as it were, in advance. When the white men arrived, they did not exactly 'flock into the settlement' – there was nothing in Monaro that could be called a settlement – but they seeped into the sheep and cattle stations, alien land now, but formerly their homeland. And there they rotted.

Disease destroyed them. It is likely that the viruses were already taking their toll of lives before the squatters took the land; but no chronicle of sickness and death was made then, or can be made now. We can, however, identify the first murderer: syphilis. Influenza was the second

---

[1] The only reference to physical conflict found in many months of search is Haygarth's reference (p. 150) to the grave of a man 'speared by the blacks' – he does not say where.
[2] W. E. H. Stanner, *After the Dreaming* (The Boyer Lectures, 1968), p. 9.

murderer and a medical historian would enumerate the others; but in the contemporary records syphilis invariably heads the list. Let us call to witness George Augustus Robinson, who kept the journal of a two-thousand miles journey that he made in 1844 through the tribal territories of south-eastern Australia. Near Omeo he wrote, 'A loathsome disease (Syphilis) among the natives, imported by Europeans, is making ravages.' Near Cooma he wrote, 'Syphilitic and other European Disease among the Natives is prevalent, and their numbers are rapidly decreasing.' Time and time again, he made entries to the same effect in his diary. At the end of his journey, he put it on record that syphilis was rife almost everywhere.[1] Reading Robinson's diary, we remember Lhotsky's tirade, ten years earlier, against the 'extinction of an entire race of men'.

Neither Lhotsky nor Robinson pointed to disease as the sole killer. According to Robinson, warring tribes were killing each other in bloody battles. Yet, in the past, it had been tribal custom to call the fighting off as soon as each side had suffered a few minor casualties; if the tribes were no longer following that good custom, something new – possibly their despairing competition for diminished resources – must have corrupted them. Robinson's imagination did not stretch to the corruption of a society; but Lhotsky's did. Lhotsky reported, not inter-tribal slaughter, but the pauperisation of the tribes. After an encounter near Gundaroo with a crowd of 'half savages' who pestered him for food and drink and tobacco, he reflected as follows: 'I was confirmed here in my observations made often in Sydney that the English had acted with these blacks as high people commonly do with the poor or beggar, to throw down a crust from their table, or a penny out of their pocket, without being willing to inquire into their real wants, or in giving them advice to remedy their poverty in a radical way by *labour*.'[2]

It was a pity that Lhotsky did not end that sentence with the words 'to inquire into their real wants'. Their deepest want, when tribal society became a wreck, was admission to the society that was supplementing it. When admission was offered to them on terms which they understood, they accepted it. They took pleasure and pride, for example, in acting as guides to the explorers – if there ever were any explorers; the persons thus designated in the white man's history books or folk lore seem almost always to have had a black man to show them the way. More often than not, the white man did not trouble to give his black guide a name, or, at any rate, to record it; thus we read of 'the native who was with us', or 'my

---

[1] *JRAHS* xxvi (1942), 318–49, for Robinson's report (ed. George Mackaness) of his journey. Robinson at that time was Chief Protector of Aborigines in Victoria.

[2] Lhotsky, *A Journey from Sydney . . .*, pp. 43–4. Chiefly, Lhotsky considered the white society responsible for pauperising the wrecked tribes; but he also blamed the government. For example, he ridiculed its practice of presenting half-moon brass plates to so-called 'chiefs' or 'kings', who received no respect from their own people.

blackfellow', or 'my tamed black'. Sometimes, however, the names of the black guides have survived in print; thus we know that Jemmy Gibber, Carbon Johnny and Boy Friday were guides at different times to McMillan, and that Charley Tara not only guided Strzelecki and his party but saved them from starvation by catching, killing and cooking 'monkeys and pheasants' – or, as we now call them, koalas and lyrebirds. Thus did Charley Tara gain admission, if only fleetingly, to the new society. The name of a beautiful river – it is a sign of grace – commemorates him.[1]

We know the names of a few Aborigines who took the same pleasure and pride in rendering the same kind of service to squatters and stockmen. For example, Tallboy helped a squatter and his men to break the law by shifting scabby sheep to fresh pastures. By devious and secret ways he guided the party through the danger zone where their crime might have been discovered; he made fire for them in a storm of rain which had ruined their tinder; in the supreme crisis of their journey, he made a bark canoe in which he and the squatter crossed the swollen Murray River to a source of supply on the south side.[2] Less melodramatic, but even more significant, is the following story, told by an intelligent Englishman, about people in a lonely hut.

> There were two black youths residing in the hut with the stockmen; we were informed that they made themselves useful in minding the sheep, milking the cows etc. The stock-keeper observed that the Blacks stopped with them better than their countrymen generally do, because they treated them more like companions, and gave them a part of such provisions as they themselves eat instead of throwing scraps to them, as if to dogs.[3]

In this sentence, 'companions' is the key word. Lhotsky, when he wrote of the 'real wants' of the Aborigines, was on the right track; but he got off the track when he suggested that 'labour' was the thing they really wanted. Labour would meet their need only on three conditions: that they should be able to find some meaning and some interest in it; that they should receive an appropriate reward for it; that they should find companionship in it. Participation, perhaps, is a better word than companionship. Men like Tallboy and Charley Tara were eager participants in the life of the new society. The invitation to become a participator, however, was not often made to an Aborigine.

[1] For the cited allusions to black guides, see J. MacKillop, 'Notes of a Journey from Sydney to Monera . . .' in *Quart. Journ. Agric.* (Edinburgh) VII (1836), 161–9; also Nos. 19, 23, 24 of *Letters from Victorian Pioneers.* For further information on Charley Tara see Brodribb, *Recollections of an Australian Squatter,* pp. 23, 28, 30, 32.

[2] 'Old Reminiscences' by 'Corio', in *The Australasian Supplement,* 13 December 1884. The writer describes in detail how Tallboy made the canoe in less than an hour.

[3] *A Narrative of A Visit to the Australian Colonies* by James Backhouse (London, 1843), p. 233. Backhouse, a Quaker of inquiring mind, stayed six years in Australia (1832–8) and kept a journal on which his book is based.

Spoiling

Understandably enough, the squatters were preoccupied with their own need for labour as an instrument of production. During the 1830s and 1840s, labour shortage in the pastoral industry became for a time matter for anxious debate in the Legislative Council and its committees. Repeatedly, the suggestion was made that the Aborigines should be used as a means of overcoming the shortage; repeatedly, the answer was given that they would never be of much use, because they invariably went walk-about after a short period of work. But might they not adapt themselves to the ways of the pastoral society if they were offered a real share of its opportunities? Some people suggested that they would work better if they were paid better. But that was an incomplete answer to the question.

The man who came closest to answering the question was Commissioner John Lambie. On 11 January 1843, in a covering letter to his report on the Aboriginal population, he referred to three Aborigines who had joined overlanding parties to Gippsland and had remained there as stockmen. 'One of them,' he wrote, 'now owns several head of cattle, which he has received as compensation for his services, and which appear to attach him to his employment; but the others seem less contented and intend to return to their tribes.' Lambie reported the following year that the man who owned cattle still remained attached to his employment and was the only Aborigine making progress towards a state of civilisation. We need not be surprised to see this man making progress. To possess livestock made a man a full participator in the pastoral life. It put his feet on the first rung of the ladder of opportunity up which, as we have seen, even a ticket-of-leave man could start climbing. The second rung of the ladder was for a man to possess land of his own – no matter what the title – on which to graze his own animals. Lambie pleaded with his masters in Sydney to give the ticket-of-leave man freedom to climb to that second rung.[1] No record has survived of Lambie making the same plea on behalf of the Aborigines.

An opportunity was missed. In Monaro, it was missed irretrievably; but elsewhere in Australia comparable opportunities are now offering themselves. This time, if they are missed, the consequences will be tragic not only for the black Australians, but also for the white Australians.

[1] See p. 43, above.

3. Improving

In the late 1830s, while the squatters were still riding the crest of their wave, Macaulay began to write his *History of England*. Looking back six generations to the glorious revolution of 1689 he felt sure that his country had ever since then been riding a wave more splendid than any that had ever before been known in any place or time. He defined this British splendour in one word, *improvement*. The word signified for him the success of a vigorous society in developing its material resources; it also signified the society's achievements of mental and moral self-development. Following Macaulay, we shall consider successively the material and the moral progress of Monaro's white people from the early 1830s to the early 1860s.

The word *improvement* was an early immigrant to New South Wales. It was the watchword of the Agricultural Society, founded in 1822 by seventy citizens of Sydney. Two years later it became the watchword of a large capitalistic venture, the Australian Agricultural Society. In Monaro, it was the watchword of Henry Haygarth.[1] In his use of it we hear intonations of nostalgia: improvement of 'the new country', it seems, means doing everything that a man can to make it look like 'the old country'. This tacit assumption contains by implication and anticipation the programmes, frequently fantastic, of the colonial Acclimatisation Societies.[2] Nevertheless Haygarth and his fellow squatters were not sentimentalists or doctrinaires, but practical men. Like the Swiss Family Robinson, they had brought with them to a strange land the varied impedimenta of their civilisation – tools, techniques, animals, seeds, slants of vision, habits of thought. These endowments, they believed, were all that they would need in their work as improvers of the land. More than a century was to go by before improving landowners were invited to entertain the idea of grazing red kangaroos companionably and profitably with their merino sheep.[3]

The native pastures, in contrast with the native animals, suited the needs of the early settlers. Only one or two individuals conceived the idea of improving them. Charles Throsby set aside 25 of his 950 acres for the

---

[1] *Recollections of Bush Life*, pp. 1, 11, 12, 14, 32, 45, 147.
[2] See pp. 115–17, below.
[3] H. J. Frith and J. H. Calaby, *Kangaroos* (F. W. Chesire, 1969). Grey kangaroos, which are predominant in Monaro, receive comparatively scant attention in this important book.

cultivation of clover and rye grass.[1] Captain W. H. Hovell cherished a larger hope, which found expression in the diary that he kept during his famous expedition with Hamilton Hume to the Port Phillip District.

> In every place where we have stopped all night, and the soil has been good, I have planted peach stones. I have also had the ground broken up with a hoe and sewn [*sic*] with clover, rye grass and burnett seed mixed . . . These I have sown in three or four places a day, as we travelled along.[2]

Had it not been for the inveterate habit the Aborigines had of setting the grass alight, Hovell would have had no fears for the success of this early experiment in pasture improvement. In the event, three generations went by before Australia's men of science brought success within sight.

Success with the breeding of animals came more quickly; but not quite so quickly as has been traditionally believed. Throughout the early decades of colonisation, cattle and sheep were wanted for their meat. All the cattle were of North-European strains; but among the sheep were some merinos of Mediterranean ancestry. Early in the second decade of settlement, John Macarthur conceived the idea of breeding merino sheep to produce fine wool; but he felt moved to complain sixteen years later that his idea was catching on very slowly. Samual Marsden was breeding sheep for meat; most of the other settlers were allowing the sheep to arrange their own breeding. As late as 1830, whale and seal oil still held precedence over wool in the list of export values. John Macarthur, nevertheless, had proved his point. 'The development of the Australian merino sheep,' it has been said, 'stood alone in the first century of settlement as a major advance at the biological level.'[3]

The sprawling squatters did nothing to speed this advance and may well have endangered it. David Waugh and the other propagandists of pastoral expansion put the whole stress on increase of numbers, irrespective of quality. Given the primitive technology of the pastoral frontier, it is hard to see how they could have done otherwise. Consider, for example, the experience of William Whittakers: it began in June 1839 when he came to terms with a Mr Ross for the purchase of a station on or near the Snowy River. This agreement invites comment on a number of counts. First, it puts no price on the land: the Crown owns the land and is willing to lease it only on an annual tenancy; for this reason the land is 'thrown in' – as common usage puts it – with the articles of current commercial value. Secondly, *improvements* – in the plural – are a well established term of commerce; but in the squattage districts they seldom amount to much.[4]

[1] T. M. Perry, *Australia's First Frontier* (M.U.P., 1963), p. 24.
[2] *JRAHS* VIII (1921), 378, Hovell's diary, 21 December 1924.
[3] C. M. Donald in D. B. Williams (ed.), *Agriculture in the Australian Economy* (S.U.P., 1967), p. 57.
[4] The agreement was dated 29 June 1839; the final contract, which allowed Whittakers a discount for cash payment, was dated 2 July 1839.

New arrivals

Agreement to purchase

	£
102 head of cattle at £4 each	408
3 horses	105
Improvements	60
Cart, etc.	30
	603
Reduction	18
	585

Whittakers gets some rough bush carpentry for £60 – just about one-tenth of what the station is costing him. Nine-tenths of his cash payment are for the cattle, horses and the cart. He is buying no sheep. He will make his start as a squatter on the cattleman's frontier.

In September 1834 a man named Jauncey had spent three or four days trying to get a mob of sheep across the Snowy River. After many false starts he hit on a bright idea.

> I said I'll be damned if I'll be beat – get the strong hide rope. I made one end fast to the tail of the dray. I said now catch 6 or 7 rams and tie one behind the other on the roap. We tied 7 (I had nine bullocks – one in shafts – pole-drays were then only just coming into vogue). I said now then dog the sheep in amongst the rams. I drew forward, the rams had to come, the sheep kept with them – not one stopped behind, there was about 16 killed and drowned, those killed got between the spokes of the wheels.[1]

Notwithstanding this exploit, cattle remained predominant in the Snowy country until the mid-1860s. Whittakers ran sheep from time to time; but like everybody else in that district he found them a risky proposition. Conspicuous among the advantages of cattle was their sheer size. The dingos could not pull them down and they could therefore be allowed to run free, with a consequential cutting of labour costs. Moreover, they could walk much further in a day than sheep and therefore saved time and money on the road to market.

Transport costs on the pastoral frontier were so high as almost to annihilate the advantage of cheap land.[2] They were a main component of the cost of living and the wages bill. In an environment unfriendly to agriculture they compelled every squatter to establish his own cultivation paddock and in other ways to seek an expensive self-sufficiency. They cut

[1] *Transcript of Notes by J. Jauncey, written about 1889 or 1890* (copy, S.M.A. Archival Records). In the early 1840s Jauncey was working for Whittakers.

[2] Space does not admit an examination of the immense transport difficulties and their economic consequences. Some suggestions for examining them will be found in Appendix 2, Tasks. The best discussion of them so far is by Lyndsay Gardiner in *Eden–Monaro to 1850* (M.A. thesis, 1951, deposited in A.N.U. Library, Canberra). This thesis was the first historical work on Monaro which I read, to my very great profit.

a large slice from his profits in the produce markets. Marketing, anyway, was a gamble, because the man on the frontier was always weeks if not months behind the news of price movements. Throughout the 1830s, the delivery of mail stopped at Goulburn; in the early 1840s, it got as far as Braidwood; in 1847, at long last, Her Majesty's mail came all the way to Cooma – but only once a week. Up to 1847, Amos Crisp of Jimenbuen used to ride the 50-odd miles to Cooma once every month in the hope of picking up mail which travellers had passed along from hand to hand. Once every year he made a longer journey.[1]

> My father at that time [William Crisp remembered] used to make butter and cheese, and once a year he would cart it to Goulburn on a bullock dray. On the return journey he would bring back supplies for his station . . .
>
> To make up his load for Goulburn all diseased cattle or very old cows would be killed and boiled down, and the fat after being rendered would be put into casks of his own making. These casks would be made by folding a cow-hide, then sewing it with strips of hide after removing the corners, leaving an opening at one end. When full a cask would weigh about two hundredweight. The beef from the boiling down would be fed to pigs, and the bacon would help to make up the load, together with hides, salted, dried and folded.

Where to find a market for the young and healthy cattle was a headache. There was no market in Monaro and to drive a mob all the way to Goulburn could take three weeks or more. Eventually, economic progress in Victoria produced the answer. In the late 1840s, Crisp and Whittakers started overlanding their cattle through the rough mountains and forests to the rich Gippsland plain. In the middle 1850s, the meat-hungry miners of Victoria provided a booming market for the cattle of widely scattered districts. The Riverina became the holding and fattening paddock for store cattle driven in from an immense hinterland; Monaro, by contrast, remained a modest provider of stores for fattening in Gippsland. In this contrast may be seen a main explanation of the Riverina's twenty years' lead over Monaro in the procession of economic development.

For men like Whittakers and Crisp there were no rich prizes; but a solid prosperity was within their grasp, provided they worked hard. Among the Whittakers papers there survives a drawing and description which A. W. Howitt made in 1868 of the home beside the Deddick River.[2] Looking down on it from the granite hills that hemmed it in, one could have mistaken it for a village – a ribbon of cottages, sheds and yards stretching the whole length of a narrowly confined promontory. Wherever there was a pocket of good soil, trees, flowers or vegetables grew. Across the river was a croquet lawn. Unless William Whittakers planted lucerne in the evening of his life on the Gippsland plain, that lawn was the only improved

---

[1] *Early History and Incidents in the Life of William Crisp* (M.L.). See pp. 107–8, below.
[2] On Howitt, whose daughter Annie married Edward Whittakers, see pp. 18–20, above.

pasture that he ever made in all his fifty-four years of life on the land.

Let us now move inwards from the cattleman's frontier to country where sheep were beginning to take first place. As Henry Haygarth rode towards this country in 1839 it seemed to him that every stage of the journey was a descent in the scale of civilisation. His feelings were mixed when he reached his destination.

> My companion asked me what I was thinking about; I would not tell him I was thinking of home. And yet I was better off than many, for we had pur-chased an 'improved station'. Scattered here and there over a considerable space of ground stood the various buildings, eight or ten in number, of which it was composed. In front was the owner's residence, a better sort of wooden cottage, chiefly distinguished on the outside by a verandah; behind and on either side of the house were several huts of an inferior structure, the abodes of working men. The wool-shed, a long rambling building, surrounded by several low sheep-yards, stood out by itself; while on a distant 'flat' appeared a large space, fenced for a wheat paddock; and in another direction a most formidable-looking enclosure, covering about half an acre of ground, formed the stockyard for cattle. The whole was backed by some low hills, thinly wooded, and agreeably receding in the distance, and at the foot of these appeared a chain of ponds or 'waterholes'. The general aspect of the place, though holding out, it must be owned, but little prospect of luxury . . . had yet, I remember, an interesting and primitive air, as it thus appeared starting up in the midst of desolation; and this it was which caused it to find favour in the eyes of its new occupants, and stimulated them to toil for its further improvement.[1]

Neither then nor later did Haygarth understand how long and hard the toil of the improving proprietor must be. In the late 1850s Mr R. Meston, J.P., a gentleman of wide experience and a highly individual prose style, conducted an official investigation into the diseases of sheep and cattle. He thus described the sheep of New South Wales – 'Bastards every one, with small frames and impaired constitutions'.[2] This, no doubt, was rough-and-ready comment on the Camden flock; but it was fair comment on the sheep in the squattages. Of course they were bastards; their owners had not yet begun to enclose them in fenced paddocks and consequently had little control over their casual matings. In Monaro, the purposeful improvement of flocks by culling and selective breeding did not get well under way until the 1870s.[3] Up to that time, the squatter kept his flocks on 'sheep stations', scattered around the property at distances usually of from 2 to 5 miles from the 'head station'. A sheep station consisted of a hut, hurdles for the folds, and a portable box large enough for a man to sleep

---

[1] *Recollections of Bush Life*, pp. 11–12.
[2] V.P. Leg. Ass., 1858–9, vol. 2, pp. 307–14, para. 34.
[3] The time lag again: in the Riverina, fencing was making good progress in the 1850s; but chiefly for cattle.

in; a team of three – two shepherds and a watchman – manned this primitive outpost, with two or three flocks in their charge. Soon after sunrise the shepherds took the flocks from the folds; throughout the day they kept them within sight; at sunset they brought them back to the folds. The hutkeeper counted them in and out and slept close to them at night in his box, trusting to the dogs to wake him if danger threatened. These tasks were not exacting and required little skill; a weaver from Manchester or a button maker from London could perform them adequately after a few months' experience.[1] They did, however, require careful superintendence by the masters and patient performance by the men. More often than not, this essential condition of flock management remained unfulfilled.

'Men of all trades, professions, degrees and callings,' Meston complained, 'jump into the management of pastoral avocations with as seeming readiness as some raw gold diggers jump into the claims of better miners. The results are too painfully manifest...The worst epidemic among Colonial livestock is the malady of mismanagement.'[2] The squatters, on the other hand, usually took it for granted that their troubles would be at an end if only they could get all the labour they wanted at a price they could afford to pay. Year by year from the middle 1830s to the early 1840s Select Committees of the Legislative Council considered ways and means of coping with the labour shortage. When the transportation of convicts was stopped in 1840 the shortage became critical. Impatient and imprudent people wanted the government to bring in coolie labour; that remedy was rejected, but no alternative remedy was discovered. In consequence, too few shepherds found themselves put in charge of too many sheep. Everybody agreed that a flock ought never to be larger than 500 ewes or 600 to 1,000 dry sheep; but some flocks in Monaro contained 2,000 sheep. If, as too often happened, one of the shepherds deserted his flock, his mate would find himself responsible for twice that number. Giving evidence on 6 July 1841 before the Select Committee, Stewart Ryrie laid heavy emphasis upon the damage thus done to the health of the sheep and to their yield of wool. At that time, the average weight of a fleece, even when conditions were favourable, was a good deal below three pounds.[3]

Good managers chose well drained slopes for their sheep stations, but bad managers allowed the men to establish them in valleys with easy access to water. Good hutkeepers planted vegetable gardens and kept the

[1] Leg. Co. of N.S.W., *Report from the Committee on Immigration 1841*, Appendix D, Sketch of a Shepherd's Duties.
[2] Meston, paras. 35–7.
[3] H. of L. Select Committee on Colonization from Ireland. Evidence of W. Bradley Esq., 8 July 1847, 3346.

living quarters orderly and clean, but bad hutkeepers turned the sheep station into a slum.

It is impossible [Meston wrote][1] to avoid denouncing the mud-holes of confined, low-situated, wet and dirty yards, which too often yet offend the eye, in spite of reason, common sense and experience. These render their due rewards, and mete out their full measures of death and destruction, of losses and crosses, of waste complete, and of ruin triumphant.

Ovine catarrh was a conspicuous dealer of death. It was a highly contagious pulmonary infection which decimated the flocks of the squatting districts throughout the 1830s and 1840s. The symptoms were large strings of mucous membrane hanging from the sheep's nose, vomiting, and such a loosening of the fleece that it could be rolled from the pink flesh of the dead animal. Catarrh was a quick and certain killer. Scab, by contrast, was not a killer; but it did even more financial damage than catarrh because of its rapid spread through the squattages by direct and indirect contagion. Its symptoms in an infected flock were biting, scratching, bared flesh, discoloration of the diminished fleeces and abrasions of the fibres. A scabby flock remained valueless unless and until the disease was cured.[2]

While these tribulations were affecting the squatters they suffered the even worse calamity of a financial slump. 'I may mention,' Haygarth wrote, 'that during my residence in Australia merino ewes have been worth two guineas each, and they have also been sold for a shilling.'[3] For three or four successive years in the 1840s improvement was ruled out; the best a squatter could hope for was to survive. Many squatters did not survive. William Bradley considered Monaro well rid of them; for the most part, as he told the House of Lords committee in 1847,[4] they were Sydney speculators who had bought stations on the rising market with a view to selling at a profit: when values fell, they collapsed. Genuine users of the land, by contrast, battled on and came through the slump as strong as, or stronger than they had been before. There had been no diminution, Bradley insisted, in the real resources of the colony.

From the *Recollections* of W. A. Brodribb a vivid picture emerges of Bradley's battle for survival. Like the other battlers he exploited the horrible expedient of slaughtering large numbers of his animals and boiling down their carcases for tallow. Of more lasting significance was his determined and carefully planned attack on the diseases infecting his

---

[1] Para. 12.

[2] These diseases were discussed extensively from the 1830s to the 1860s in pamphlets, books and parliamentary papers. The Scab Act of 1853 (12 Vic. No. 27) is the first important landmark of legislation to protect and promote animal health in New South Wales.

[3] *Recollections of an Australian Squatter*, p. 161. Brodribb's figure was 1s. 6d.

[4] 3383–3390.

flocks. His remedy for catarrh was short – immediate slaughter of the infected animals and removal of those still healthy from wet and cold ground to a dry hillside facing the morning sun. To rid his flocks of scab he worked out an elaborate drill, practised it himself and then demonstrated it to his managers. In this drill the preparatory moves were to erect three or four bails in a small yard and to prepare large quantities of a medicinal mixture – 1 lb of tobacco leaf to 1 gallon of water: the mixture was ready after it had been boiled for three quarters of an hour: supplies sufficient for the day's work were kept on hand. When these preparations were complete the sheep were washed, sheared and brought in successive batches to the bails: there, two men examined every sheep and scarified every scab with a knife or small curry comb: two other men dipped and soaked every sheep for four minutes in the hot tobacco water: two other men plunged the sheep into a tub of clear water and then rubbed off the residue of their scabs with a wooden hoop. That completed the drill. Taught by Bradley, Brodribb supervised every motion of it throughout the entire day. A working day's tally was 150 sheep. The work continued from day to day until not one scabby sheep remained in the flock. Thus did Bradley make clean 300,000 acres of Monaro; thus, later on, did the Chief Inspector of Stock make clean every property in every district of New South Wales. Victoria, meanwhile, remained a scabby colony.[1]

Bradley was at need a ruthless manager of his flocks; but he was invariably a gentle manager of his pastures. He knew no more than the other squatters did about the native grasses; he made few experiments with exotic species; what he did preach and practice was prudence in putting stock on the pastures. Even on the very best land, he told the House of Lords committee, it was imprudent to run more than 1 sheep to 3 acres; since every property contained some admixture of better and worse soils and grasses, the average stocking rate ought to be *lower* than 1 to 3.[2] Just as his aggressiveness helped him to survive the slumps, so did his self-restraint help him to survive the droughts. Moreover, he was in all situations a careful calculator of financial cost and profit. In the Brodribb narrative he stands revealed as a pathfinder in the slow advance of pastoral Australia from the ill-considered exploitation of natural resources towards the still distant achievements of scientific conservation and development.

Conspicuous among the landmarks of economic recovery in short term were the commercial upswing of the late 1840s and the gold discoveries of the early 1850s. Towards the close of the latter decade, Monaro had its

[1] The mixture prescribed by the Chief Inspector of Stock contained less tobacco leaf than Bradley's mixture but made the deficiency good with sulphur. See Alexander Bruce, *Scab in Sheep and its Cure* (Sydney, 1867).

[2] 3342. Cf. 3339–40, giving Bradley's own figure of 40,000 sheep to 300,000 acres (= 1 to 7½). But of course he was running cattle too.

own lucky strike; but the short-lived Kiandra boom has no place in this story of land use, except for the additional impetus it gave to the already accelerating improvement of transport, markets, industrial and commercial services, and other institutional features of the economic environment. More specific in its consequences for land use was the decision taken on 9 March 1847 to convert the squatters' annual tenancy into a fourteen years' tenancy. That decision had emerged from conflict; it contained within itself the seeds of future conflict; but W. A. Brodribb correctly reported its significance as the squatters saw it. Hitherto, he wrote, they had had 'no inducement to improve'; henceforward they had the inducement 'to erect permanent improvements on their squattages'.[1]

A precise enumeration of the improvements erected on William Bradley's twenty stations in Monaro appeared in an advertisement of sale which the *Sydney Morning Herald* published on 1 May 1866.[2] For Coolringdon, formerly the home of W. A. Brodribb and still the administrative centre of Bradley's land, the main items enumerated were as follows: a stone house of ten rooms with verandah and large cellars, galvanised iron roof and water laid on to the kitchen, bathroom and garden: a detached stone kitchen and scullery with two rooms for servants, a bathroom, an office and a large nursery; a spacious store, newly built of wood and roofed with corrugated iron; a stone cottage of five rooms with a garden in front; eight huts for workers on the station; stables, coach house, hay loft, limehouse, fowl house, and a blacksmith's shop complete with tools; two lucerne paddocks totalling 6 acres; four cultivation paddocks totalling approximately 40 acres; two grazing paddocks enclosing about 300 acres; a woolshed large enough to hold 1,000 sheep under cover; a detached stone wool room 40 ft. × 16 ft.; drafting yards; a stockyard; a covered milking pen and a calf pen. All these improvements were contained within 1,300 acres of freehold bought by Bradley during the past nineteen years. That acreage was not quite one-twelfth of the total; by and large, Coolringdon was still an unfenced leasehold property. On its twelve out-stations conditions were comfortable and seemly for the men and the animals. There was a new and well-equipped washpool on the river; but there were no fences and few if any innovations in the technology and routine of pastoral practice. The same was true of Myalla, Dangelong and the other Bradley stations. Their areas of freehold and the scale of their improvements were less imposing; but masters and men had housing appropriate to their situations in the pastoral hierarchy, the station equipment was in good order and the animals in good condition – yet still the traditional routine persisted. Westwards in the Riverina and eastwards on the coastal plain the pace of change had been faster. Even so, Bradley had good reason

---

[1] *Recollections of an Australian Squatter*, pp. 53, 57.
[2] See p. 91, below.

to feel that his four decades of strenuous effort had produced some improvement in Monaro.

Other men, who possessed far fewer of this world's goods than Bradley did, had to their credit achievements comparable with his, in quality if not in quantity. Among them was John Boucher of Bukalong, a homestead charmingly situated in gently undulating country ten miles north-west of Bombala. In newspaper reports from the 1840s to the 1880s John Boucher's activities received frequent attention: he was a Justice of the Peace; he nominated sound conservative candidates for the parliamentary elections; he collected subscriptions to help the Mother Country win her war with Russia; he called a meeting to raise money for a statue to commemorate the late Prince Consort; he donated £100 to the endowment fund of the Diocese of Goulburn.[1] All these good deeds made news; but no journalist ever reported the supremely good deed for which we today still bless the name of John Boucher. In the new year of 1858 he started to keep rainfall records at Bukalong. For the next 26 years he kept them day by day.

In 1884 Bukalong passed into the possession of the Garnock family. Their connection with Monaro had begun in 1835, when George Garnock of Glasgow squatted with his sheep a mile and a half from Bukalong – at Mount Misery, which he renamed Mount Pleasant. George Garnock had already gained colonial experience as supervisor for two of the leading pastoral families, the Brooks of Denham Court and the Macarthurs of Camden; when he came to Monaro he came to stay. His grandchildren and great-grandchildren remain to this day deeply rooted in the Bombala district. At Bukalong they have faithfully performed the task to which John Boucher set his hand. From 1858 up to now, not a single point of rain falling on Bukalong has gone unrecorded. In the whole of Australia only Adelaide has provided a continuous record of longer duration.

During the 1850s, the botanist von Mueller and the geologist Clarke gave a great lift to scientific discovery in Monaro; but their incursions were brief and discontinuous. For the advancement of science in long term, the regular recording of long-continuing observations was also required. That was John Boucher's contribution. One wonders into which of his two categories Macaulay would have put it – material improvement, or mental and moral improvement. What follows will be a brief consideration of a few questions which belong conspicuously to the second category.

The mental and moral improvement of white society in Monaro had to begin with a climb out of a pit. How deep and dark the pit was, Lhotsky

[1] Perkins, 223, 387, 431, 455, 537, 744, 746.

discovered in 1834. Like Haygarth after him, he saw on his journey south the outward and visible signs of a descent to barbarism – no church south of Sutton Forest, no window pane south of Canberra, no white woman south of Micalago. For him, much more than for Haygarth, these superficial observations were pointers to the condition of society. At Micalago he found himself at 'the limits of location'. The writ of government ran no further. 'I had lived before,' he reflected, 'under absolute monarchies and under commonwealths; here I found myself surrounded by absolute anarchy.' Anarchy, he believed, spelt degradation. There was a sly grog shop at Micalago. He thought it was a den of thieves.

A few miles beyond Micalago he had a melancholy encounter.

> In this lonely place we were met by a prisoner, belonging to a neighbouring station, who barefooted and covered with rags, reminded me forcibly that I was in a land of banishment and expiation. I asked him how he came to be so badly off, he replied that the slops were issued very irregularly, and was besides of the worst description. He was also all over affected with the syphilitic disorder, and told me, that many men were in the same situation, without any surgeon at hand. I shook my head, as it appeared to me, as if some demon sentenced to perdition was addressing me in this valley of desolation.

In an indignant footnote he attacked the social and political system which bred such vice and misery.[1]

Lhotsky could not foresee that absolute anarchy in Monaro had only five more years and the transportation of convicts to New South Wales only six more years to run. He was making his journey in the dark before the dawn. Even in that dark, he saw gleams of hope; he met masters who were not callous and men who were not depraved. Some shepherds were living in filth, but others were caring for their huts and making good gardens. Self-respecting or shiftless, healthy or sick, they all received him with generous hospitality. With some of them his talk around the fire took surprising turns. On one occasion some savage-looking men who had frightened him at first expressed 'much British pride' when he described to them the glories of the Zoological Gardens, the British Museum and other institutions of their Mother Country.

He was right in his refusal to regard with despair the future of the convict population. In the late 1830s and early 1840s a young Englishman of good education, James Demarr, spent five years on a self-conducted tour of the eastern colonies, always at the working class level. He enjoyed his adventures but did not romanticise them. From time to time he met the kind of man who would not hesitate, as he put it, to sharpen his knife on his father's gravestone to cut his mother's throat; but when he fell ill the good Samaritans who succoured him were convicts. By and large, his prognosis for the pastoral labour force was hopeful. The climate was good,

---

[1] Lhotsky, *A Journey from Sydney* . . . , pp. 79 ff.

the food was ample. More important still, a man had a chance to show his mettle; a Londoner or a Lancashire lad could soon turn himself into a first-rate stockman, even though he had never ridden a horse until he was lagged to New South Wales. Free or unfree, a man could possess his own dog. Demarr felt convinced that the upward pull to decency and self-respect was stronger than the downward pull to degradation.[1]

Haygarth felt the same conviction. He described as follows what seemed to him to be one of the most remarkable sights of Australia – 'the man who, having been rejected by the place of his birth, and of his early crimes, has paid the penalty, has passed the period of his disgrace, and has returned to a better life in another land'.[2] When he wrote those words servile labour was moving towards the end of its long lease of life. The census of 1851[3] enumerated seventeen ticket-of-leave men in the Bombala district and five in the Cooma district; apart from that tiny minority freedom was now 'the civil condition' of everybody in Monaro. This freedom contained various categories; convicts whose sentences had expired or who had received pardons; currency lads and lasses who had been born free: immigrants who had come to New South Wales of their own free will. Contemporary evidence strongly suggests that the rate of crime in this diversely constituted population was not abnormally high: as one would expect, it was higher among the former convicts than among the free immigrants: as one would not expect, it was lowest of all among the native-born Australians.[4] Since most of the last named were of convict parentage, some interesting questions arise – among them, questions about the influences of a good climate and a buoyant market for labour.

A close examination of the police court records of Cooma might produce some answers to those questions.[5] In 1854, the most frequent offences against the law were as follows: obscene and insulting language; drunkenness; assault; absconding from a master; theft of livestock or money; receiving stolen goods. Nearly all the offenders were poor people; nearly all the offences were petty. Further research might identify the offenders under the three categories listed above. It might also make possible a comparison of crime rates, then and now. Meanwhile, there seems no reason for believing that Monaro in the 1850s was irredeemably wicked.

It was suggested just now that a buoyant market for labour helped to hoist white society out of the horrible pit described by Lhotsky, but during

---

[1] James Demarr, *Adventures in Australia Fifty Years Ago* (London, 1893), *passim*. This book is good evidence because it is based on a diary, memoranda and letters which Demarr wrote in the years 1839–44.
[2] *Recollections of Bush Life*, p. 151.
[3] Table 11.
[4] The evidence is discussed by Russell Ward in his inaugural lecture, *Uses of History* (Univ. of New England, 1968).
[5] These records, continuous from 1854 to 1865, are in the Mitchell Library.

the squatting period that suggestion was seldom if ever made: after all, nearly all the people who put pen to paper were employers of labour. It would have been surprising if they had seen high wages as a main path towards moral improvement; it is not surprising that they should put much of their trust, as Haygarth did, in the ministrations of religion. We must give them credit for sincerity. For some of them – for example, for William and Louisa Whittakers – separation from the means of grace had been the occasion of deep distress. Even an irreligious family, we may imagine, would welcome the arrival of an intelligent parson with news of the outside world. The Rev. Edward Gifford Pryce, M.A., was such a parson. In 1843 he began what he called his 'itinerating ministry' in Monaro. In a report next year to Bishop Broughton he thus described his way of life: 'Having no residence, and consequently no home, I have been obliged to go from house to house; and during the whole time I have been in this district, I have not at any time been more than three days at once in the same place, except on one occasion when I was confined for a week with a violent cold.'[1] He was bound to be perpetually on the move because his parish was coterminous with Lambie's still untruncated Squattage District: from Micalago to the mouth of the Snowy River he was a familiar sight, riding along the rough bush tracks, leading his pack horse and attended by a young man whom everybody called 'the parson's Johnny'. From one point of view he thought himself fortunate in having no home, because he was thereby constrained to fulfil his mission of bringing religion into the homes of the people; from another point of view he thought himself unfortunate, because study was a duty of the clerical life which he could not possibly perform. In the late 1840s, a parsonage and a church were built for him at Cooma.

The idea caught on. In all parts of Monaro all sorts and conditions of people gave money to build churches. Most evocative of the spirit of that time is the tiny stone church of Gegedzerick station. In hallowed ground close by, the rude forefathers of Gegedzerick sleep.[2]

Haygarth believed that Monaro needed churches, but needed homes even more: the time was past, he insisted, when a man could come to the colony in the hope of getting rich and then getting out: if he came at all, he must come prepared to make a home. For most men, that meant marriage.

It is usually remarked [he wrote] . . . that single men are apt to neglect their affairs, being glad to avail themselves of every pretext for leaving home in quest of society, or, if they remain there, they are often driven to seek solace in intemperance; and the usual practical advice given to a young man, as

[1] *Account of the Mission of Monaro* by the Rev. E. G. Pryce, addressed the Lord Bishop of Australia, in Broughton, *The Church in Australia* (London, 1846).
[2] Anon., *Berridale: A History of the Parish* (1935).

soon as he gets a little settled and 'sees his way before him', is to take a wife as soon as possible . . . I have usually remarked that . . . the most contented, and usually the most prosperous settlers, must be looked for among those who in woman's love have found a balm for disappointment, and the noblest stimulus to exertion.[1]

When Haygarth wrote that passage, the absolute dearth of women which Lhotsky had reported was already a disappearing evil. Owing to the shortcomings of the early censuses, it would be a waste of time to attempt a statistical demonstration of the moving ratio between the sexes; but broadly it would be true to say that the climb of the female population was rapid from the late 1830s onwards. The Census Report of 1856 contained the following summary:

	Males	Females	Total
Cooma Police District	1,056	673	1,729
Bombala Police District	613	367	980

There was still a long distance to go before an even balance between the sexes would be achieved: even so, the figures demonstrate a demographic improvement which Lhotsky would have found astonishing. Since the female population was almost entirely in the younger age groups. Monaro could look forward to large crops of children. In bringing them into the world women would more often than not have to help each other: in the whole of Monaro, a territory closely comparable in size with Wales, there were at mid-century only two medical men. Cooma got its first government school in 1863, but did not get its first hospital until 1867.

The widely scattered community thus remained heavily dependent upon the virtues of individual self-help and mutual help among neighbours. Those virtues were not in short supply.

[1] *Recollections of Bush Life*, p. 154.

PART III

POSSESSING THE LAND

1. Battles for possession

In the mid-1840s the overmighty squatters had fought Governor Gipps and beaten him. Twenty years later they still held possession of the land.

	Stations
Over 30,000 acres	23
20,000 to 30,000 acres	31
10,000 to 20,000 acres	95
Below 10,000 acres	58
	207[1]

The names of the stations listed in 1866 were almost invariably identical with the names that had been listed in 1848. Admittedly, the names of the persons in possession were not comparably identical; throughout the two previous decades many squatters had been often on the move – in, out and round about. Even so, more than one-third of the original family names could still be seen on the squatting map of Monaro. A casual observer might have concluded that the Monaro squatters were securely dug in. Yet the opposite was true. In 1861, a democratically elected parliament had served notice on all squatters that they would have to fight for their great possessions.

By passing the Crown Lands Alienation Act and the Crown Lands Occupation Act, parliament proclaimed its determination to make room on the land for a sturdy yeomanry.[2] The more thoughtful squatters agreed that there was land enough to spare for meeting the real needs both of large graziers and of small farmers. Their spokesmen in parliament and the press called for a clear-cut demarcation of zones on the squattages: one zone reserved for pastoral tenants of the Crown, the other reserved for freeholders. This proposal was compatible both with social justice and economic common sense; but a large staff of competent surveyors would

---

[1] The Pastoral District of Monaro covered far more ground than that of the three-county area, which is the object of this study (see map 7, p. 47). Consequently, the table above inflates the number of Monaro stations. At the same time, it deflates their average size, because the stations on the coastal plain tended to be smaller than those on the tableland. The table has been compiled from the list in *Baillière's New South Wales Gazetteer and Road Guide* (Sydney, 1866). This list was based on official returns published periodically. See e.g. V.P. Leg. Co., 1854, vol. 2, and V.P. Leg. Ass., 1859–60, vol. 3.

[2] 25 Vic. nos. I and II. The parliament which passed these Acts was the first to be elected in New South Wales under manhood suffrage; but little more than one quarter of the registered electors voted. See *NSW Statistical Register for 1861*, pp. 61–3.

have been required to make it workable. That condition was not met. As late as the 1870s, New South Wales was still suffering a self-inflicted starvation of surveyors. 'I do not think,' Surveyor-General P. F. Adams declared, 'that any other British colony is in so backward a state.'[1]

In the opinion of the Surveyor-General, the land legislation of 1861 had been 'a strong remedy for a great evil': since survey before settlement was unattainable, free selection before survey was the only means that he could see of unlocking the lands.[2] Under the Crown Lands Alienation Act, any man could select anywhere on any pastoral lease a block of land as small as 40 acres or as large as (but no larger than) 320 acres.[3] The purchase price of every block thus selected was fixed at £1 per acre. The selector must pay one-quarter of the price immediately. At the end of three years, two options were open to him: either to complete the purchase, or to pay interest on the outstanding balance at the rate of 3 per cent per annum. If he chose the former option, he received title to the land in fee simple; if he chose the latter, he remained in secure possession under the title named conditional purchase. Either way, he must declare above his signature that he had fulfilled two essential conditions: first, that he had been resident on the land throughout the three-year period; secondly, that he had improved it to the value of £1 per acre. For a small man anxious to rise in the world, it paid best to complete the purchase, because he could then select another block. Moreover, he could secure with it a grazing 'right' or 'prelease' of three times that acreage. Thus did free selection provide for land-hungry people, or seem to provide for them, a splendid ladder of opportunity.

In 1872, Surveyor-General P. F. Adams declared that free selection, barring a few accidents and errors, had been 'a great success'. In 1883, Royal Commissioners Augustus Morris and George Ranken declared that it had been a disaster.[4] Their main indictment against Sir John Robertson's Land Acts was that they had offered for sale to one class of occupants the same land as was assigned under lease to another class. The inevitable consequence, they maintained, was an embittering struggle for ownership. The squattage still remained a property recognised by law; but under the new law it could be obliterated. The squatter saw all his land at risk and fought for every acre of it. He and the selector were equals under the law, but in the battle for survival he was the strong man armed: he could have blocks of land measured and put to auction; he could buy land that he had

[1] V.P. Leg. Ass., 1872, vol. 2, *Minutes of Evidence taken before the Select Committee on the Land Law*, Appendix, pp. 190–3.
[2] *Ibid.* Q. 378. On the politics of free selection see the illuminating article by D. W. A. Baker, 'The Origins of Robertson's Land Acts', reprinted in *Hist. Stud., Selected Articles*, first series, (M.U.P., 1964), pp. 103–26.
[3] The Amending Act of 1875 raised the maximum area of a selection to 640 acres.
[4] V.P. Leg. Ass., 1883, vol. 2, pp. 77–248.

improved; he could go into the free selection business both under his own name and under the names of his sisters and cousins and aunts and toddling children and submissive servants;[1] above all, as a man of substance and proved performance, he could get backing from the banks or the pastoral finance companies. In this unequal contest the squatters could not fail to win; but they won a Pyrrhic victory. 'Monied institutions,' the Commissioners asserted, 'have emptied their safes in buying up the Crown Lands on stations, and the largest and best of these properties have become little more than the assets of financing firms.'

'History,' somebody has said, 'does not repeat itself. Historians repeat each other.' Three generations of Australian historians repeated the Morris–Ranken version of the free selection story. At last, in the mid-1960s, a young historian asked himself whether or not the Royal Commissioners had told a true story of free selection in one important district, the Riverina. He thought it suspicious that they had excused themselves – unlike their predecessors, the parliamentary committees of the 1860s and 1870s – from publishing minutes of evidence and the names of witnesses. With an open mind, he checked their forceful generalisations against the evidence of the Conditional Purchase Registers and such other contemporary material as he could track down. From these researches it became quite clear that Morris and Ranken had made a biased selection of evidence. Their story of disaster was close to the facts of life around Deniliquin, but remote from the facts of life around Albury and Wagga Wagga. In the latter neighbourhoods free selection, bountiful rain and good access to markets had made independence on the land a reality for many farming families.[2]

In their discussion of the Monaro evidence, Morris and Ranken were not so conspicuously biased; but they were in too great a hurry. Two printed pages seemed to them space enough for telling the true story of free selection both on the Monaro tableland and on the adjacent coastal plain. We need to take a closer look than that. So let us now, as was agreed earlier, focus our attention on William Bradley's territory. It is a sufficient sample – twenty stations, with a combined area of above 300,000 acres. These stations stretched in a closely linked chain almost the whole distance from Cooma to Bombala. In 1866, just two years before he died, Bradley offered them all for sale. The advertisement of sale, issued in Sydney by the firm of Richardson and Wrench, throws light on Bradley's method of estate management: control of the northern stations by a manager at Coolringdon, with assistant managers at Myalla and Dangelong; control of the southern stations by a manager at Bibbenluke with an assis-

[1] 'Dummying' was the word commonly used to characterise these various expedients of vicarious selection. The Amending Act of 1875 made it illegal.
[2] G. L. Buxton, *The Riverina, 1861–1891* (M.U.P., 1967), *passim*.

tant manager at Maffra.[1] In the event, Bradley sold the northern stations only. What the buyer received was not, of course, the freehold, which the seller did not possess, except in some carefully selected pockets; he received the livestock, the improvements and a lease of the pasture. The value of his purchase would depend in large measure upon his success or failure in learning to live with free selection.

The main buyer was Hugh Wallace, a Scot from Dumfries who had built up a sound freehold property within 'the limits of location' at Nithsdale near Braidwood. Hugh Wallace wanted Bradley land for his three sons; he put William on Coolringdon, Henry on Dangelong and John on Myalla. But the Wallace family did not stick. The last of its failures occurred in 1877 at Dangelong, which by then was heavily in debt to the Sydney firm of George King and Company. The head of the firm began to feel anxiety for the safety of his investment and sent one of his sons to inspect Dangelong. Wallace had far fewer sheep on the property than he had stated; but he had good cover close to his yard, so he instructed his shepherds to keep driving the same sheep into the yard, then from the yard to the cover, then back again to the yard. The trick would have succeeded if Wallace could have kept quiet about it, but he boasted about it the same evening at a bar in Cooma. The story was passed around and came to the ears of George King and Company. They ordered a recount of the sheep. Foreclosure followed.[2]

In the pages which follow we shall get to know two men of a different stamp – men with the stamina and the wit to take and hold their land in a firm grip. The first of them, James Litchfield, belonged to a farming family in Essex and had come to New South Wales in 1852 with good experience and good recommendations. After spending a few months in Sydney and Goulburn he took service with William Bradley as manager of Myalla. When the Robertson Acts came into operation in the New Year of 1862, he saw his opportunity of graduating from management to ownership. On 8 April he made his first selection, a block of 320 acres on Jilla-matong Creek.[3] Like Brodribb before him, he received encouragement and tangible aid from William Bradley. This backing, and his own character, would make him a sound man in the eyes of the Commercial Banking Company, which had recently opened its doors in Cooma. Thus supported, he persisted throughout the 1860s in making frequent and well-planned

[1] *S.M.H.*, 1 May 1866. This very informative advertisement runs to a column and a half. It calls the northern stations *Lot 1*, the southern stations *Lot 2*, and emphasises the unity of each lot as a going concern. As map 9 shows, the only break in geographical continuity between the two lots occurs near Maffra.

[2] This story was told to me by Mr George King of Bungarby, a grandson of the original George King. Various entries in the diary of Edward Pratt (see below) incline me to accept the story as authentic.

[3] Perkins, 726.

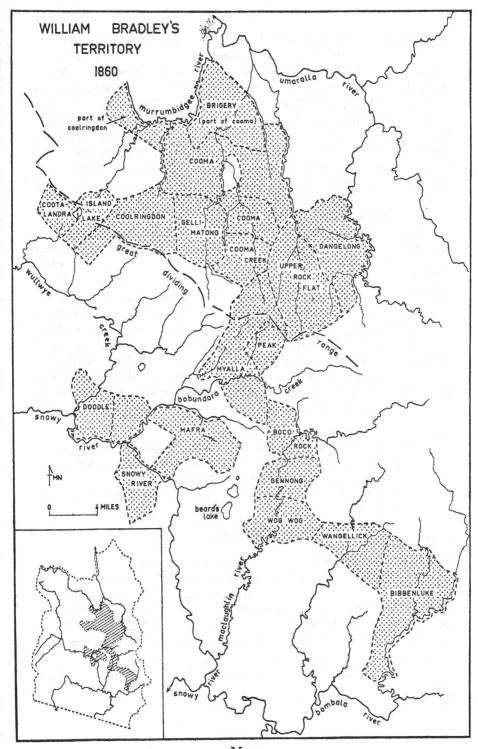

WILLIAM BRADLEY'S
TERRITORY
1860

part of
coolringdon

murrumbidgee river

umaralla river

BRIGERY
(part of cooma)

COOMA

COOTA   ISLAND
LANDRA   LAKE   COOLRINGDON   GELLI-   COOMA
MATONG   DANGELONG

great   COOMA
                                CREEK
                                        UPPER
dividing   ROCK
                                        FLAT
wullwye

creek   PEAK   range

                                MYALLA

                        bobundara   creek

snowy   DOODLE

                        MAFRA   BOCO
river   ROCK

                                        GENNONG
MN
                SNOWY
0   4 MILES   RIVER
                                WOG WOG
                        beards   WANGELLICK
                        lake

                                                BIBBENLUKE

                        maclaughlin river

                snowy river

                                bombala river

Map 9

selections, fitting the pieces together to form a substantial freehold property. On that property he made his home.[1]

With our retrospective awareness of the prominence which James Litchfield ultimately achieved in pastoral Monaro, we may be tempted to place him in the squatter class; but to do so would be wrong. He was a member, and on important occasions the spokesman, of the Free Selectors' Association of Cooma. As we have seen, he was better placed than were most of his neighbours for making a start in the long climb up the ladder of free selection; but he also proved himself a good finisher. We want to find out what his aims and methods were. In evidence which he tendered in 1874 to the Select Committee on the Administration of the Land Laws, he stated them with crystal clarity.[2]

Throughout his statement he insisted that the land user in Monaro must make his living for the most part by depasturing animals, not by tilling the soil. The minimum size for an efficient pastoral freehold, he further insisted, was 2,560 acres. That figure was not chosen at haphazard; it was precisely eight times the maximum acreage allowed for a selection under the Crown Lands Alienation Act. To be sure, the Act allowed the selector to claim a 'grass right' three times the area of his selection; but this right, Litchfield said, was fictional; almost invariably, the original selector found his grass right invaded and annihilated by rival selectors. In justifying that assertion, Litchfield faced some probing questions. Members of the select committee wanted to find out whether the invading selectors were acting in good faith on their own behalf, or acting as dummies for somebody else.

1938. In whose interest? We could not tell.
1939. Did the parties generally take possession? Shepherds have generally taken selections out of the grass rights, and we cannot tell whether they are for themselves or for the lessee of the run.
1940. Are those selections generally improved and resided upon? No – forfeited.
1941. Are they taken with that intention? Yes.
1942. To drive you off the grass right? Yes.

The richer man was able to drive the poorer man off his grass right because the law allowed him to bid at auction for a forfeited selection.

Nevertheless, a selector who commanded middling financial resources, as James Litchfield did, possessed some useful weapons of self-defence and counter-attack. Occasionally, he might himself bid successfully at auction for a forfeited selection.[3] More commonly, he built up his freehold

[1] See W. Davis Wright, *Canberra* (Sydney, 1923), p. 50. Wright, who had been sheep manager for Bradley on Coolringdon, says of James Litchfield and his wife, 'Nothing could keep them back'.

[2] V.P. Leg. Ass., 1873–4, vol. 3, pp. 995 ff.

[3] *Ibid.* 1845.

by making selections in the names of his children. This practice, it was put to Litchfield, was an evasion of the law; but he maintained that the law not only permitted it but also – if the law's intentions were not to be frustrated – required it. A member of the committee then asked him to explain how a child could fulfil the condition of residence; to explain, in particular, how his little daughter had resided on a selection registered in her name.

1871. How was the residence fulfilled? Her residence was not continual.
1872. How did she perform the act of residence? By going on the ground occasionally.
1873. What do you mean by occasionally? Perhaps once or twice in a month.
1874. Did she stop there all night? Yes, that is the way residence for minors has hitherto been fulfilled, and that has been accepted.
1875. Is your daughter at school? She is at home; her mother is her teacher.

. . . .

1896. Do you think there are many cases similar to your daughter's case? I should say about eighteen out of twenty.

Under further questioning Litchfield admitted that he had been using the names not only of his children but also of adult persons in his employment. One of these persons was his brother-in-law, who was working on his land and living in his house.

1929. Did he take it [the selection] in his own interest or in yours? In mine I believe.
1930. Was there an arrangement that he should give it up to you? Yes, at the expiration of three years, if I required it.
1931. *Chairman.* That was very much like a dummy, was it not? We have been compelled to select that way to secure the grass right.

For a man to use a dummy, Litchfield maintained, was wicked, if he did it to obstruct settlement; but righteous, if he did it to promote settlement. As the test and measure of *bona fide* settlement he proposed a close scrutiny of the money and effort expended on improving the land. He cited the improvements which he himself had made as proof that his expenditures not merely satisfied, but far exceeded the requirements of the law. He expressed dislike of the subterfuges which he and his kind had been practising; but they would remain a necessary evil, he declared, until such time as the law was made conformable with the environmental and economic facts of life on the land. The law should be amended to serve a single purpose: that the selector's holding should be large enough to provide a decent family living.

1954. *Chairman.* You think he ought to be allowed to take up more land? Yes, and to do away with selection by minors. I would allow the parents to select up to eight sections, or 2,560 acres, and not permit them to make

another selection for ten or twelve years, to prevent monopoly; and
not to sell or transfer to any person until the expiration of the said term
or terms, which would prevent dummyism. I am of opinion that this
system would encourage the *bona fide* settler.

He was absolutely right; but his conception of economic viability – to
quote the phrase of a later time – was too realistic to win acceptance in the
1870s.

The amending Act of 1875 fixed 640 acres as the maximum size of a
selection. That was double the previous maximum, but in the judgment
of James Litchfield it was still 1,960 acres too small to meet the needs of
selectors in Monaro. The Act also prohibited dummying and selection
in the names of children under sixteen. Nevertheless, a resolute and
resourceful selector could still find ways and means of acquiring land in
freehold. By the mid-1880s the Litchfield family held secure possession
of 20,000 acres.[1] Some other selectors, after starting their climb lower
down on the ladder of ownership, were now secure on its middle rungs.
Among them were two shepherds, James Thompson and James Devereux.
Thompson acquired a substantial property on Ironmungie, a station which
filled the gap between the northern and southern divisions of the Bradley
territory. Devereux did even better for himself at Rock Flat on the Cooma–
Nimmitabel road. There he proved himself an enterprising flock master,
grazing his fine-woolled Merinos around Lhotsky's mineral spring. One
feels surprise at men so poorly endowed with this world's goods achieving
such conspicuous success, until one calls to mind their bountiful crops of
sons and daughters, all earning money and putting it into the family pool.

Historians, it would appear, have too much emphasised the stratification
of classes within the pastoral economy. They ought also to explore the
movement of individuals across the class barriers.[2] For us, this exploration
can be no more than incidental; our immediate task is to identify the
inheritors of William Bradley's land. But let us change our angle of
approach. After looking at men who were climbing up the ladder, it will
be interesting to look at a man on its topmost rung. Our man, Edward

[1] All the Litchfield properties in this period are listed in two superbly bound volumes which
are preserved at Hazeldean. The lists are accompanied by plans, well drawn in Indian ink on
parchment by Balmain and other surveyors. Ownership in 1884 was as follows:

	Acres
Selections	9,138
Purchases of freehold from the Crown	3,780
Purchases of freehold from other selectors	1,970
Purchase of C.P.s	5,556
Total	20,444

[2] Mr L. F. Fitzharding has shown that there was plenty of social mobility on the Limestone
Plains. See his article, 'Old Canberra and District, 1820–1910', in *Canberra–A Nation's Capital*,
ed. H. L. White (Sydney, 1954), and his edition of Samuel Schumack's *Autobiography* (A.N.U.
Press, 1969).

Pratt, bought the pastoral lease of Myalla in 1872 and found himself thereafter hard put to defend his purchase. In his diary, he painted a living
picture of his time and place, and of himself.[1]

Edward Pratt was born in a vicarage of County Cork, grew up in Ply-
mouth and went to Cambridge to read the mathematical tripos; on the
class list of 1854 he was ninth wrangler. In 1857, at the age of twenty-
six, he migrated to New South Wales to become senior mathematics
master at Sydney Grammar School. The school was lucky to get him, for
he was not only a good scholar but also a careful and thoughtful teacher
– too thoughtful, perhaps, for the taste of the boys and their parents. He
hated flummery and held unorthodox views about the altar and the throne:
scepticism, he reflected, was no bar to Holy Orders, provided the sceptic
was also a hypocrite: at divine service one Sunday morning he took note
that the Prince of Wales, not God, received the praise.[2] He remained a
taster of sermons but chose John Stuart Mill as his mental and moral
tutor: 'I owe more to that man,' he wrote, 'than any man living. Political
Economy opened a new life for me.'[3] He studied the economic treatises
of Cairns and Jevons as they appeared in print. For relaxation, he read
history books and serious novels; among his favourite authors were
Motley, Lecky and George Eliot. He subscribed to the *Spectator* and the
Contemporary Review and read them from cover to cover as soon as they
arrived from England. Throughout his life he retained his fervour for
self-improvement and the improvement of everybody close to him – his
own children, to start with. We see him in his diary as the stern Victorian
parent. Yet he was also a loving father. Year by year he visited the grave
of his little boy Theo and tended the rose tree that he had planted there.
He took a cutting from it and planted it in his garden at Myalla.

He loved England, but his roots sank deep into Australian soil. From
the start, he enjoyed the out-of-doors life around Sydney; he sailed a boat
in the harbour, went swimming in the ocean surf and grilled chops on the
beaches. With equal zest he attended political meetings and studied the
parliamentary debates. Political economy came to life for him in the
statistics of the colony's progress – in particular, its pastoral progress.
The idea took root in his mind that he would do well to sell his mining
shares and put the proceeds into a sheep station. In March 1872 he bought
Myalla.

Seen from one point of view, the timing of his purchase was good, for
in the early 1870s both the seasons and the wool prices were improving.

[1] I am indebted to Mr Gordon Ferguson, a grandson of Edward Pratt, for permission to study
this illuminating diary. A photostat copy of it has been lodged, under restricted access, in the
National Library, Canberra. When it is directly cited in the following pages, the date of entry
will be noted.
[2] 17 May 1871, 8 May 1872.
[3] 12 May 1873.

Possessing the land

Seen from another point of view, the timing was bad, for free selection was getting away to a gallop. Between 1862 and 1870, alienation of pastoral leasehold in the neighbourhood of Cooma had got nowhere near one-tenth of the total area; but a good deal more than half the total was alienated between 1871 and 1880. After the mid-1880s, only a tiny residue of the leasehold remained unalienated.[1]

When Pratt bought Myalla, he bought himself into a war. He might well have called it 'the unnecessary war'.

> I have hit on a bright idea [he wrote] for an improvement of the Land Laws. Whenever a selection is taken on any run let the squatter have the privilege of purchasing an equal area of some part of the run – give him (say) a lease of 21 years of that portion. By this means the squatter would be secure of half his run – the 'poor man' would have half the pastoral districts open to him and neither party would have an unfair advantage. The selector would have first choice in every instance.[2]

Four years later, he was still expounding similar views in letters to the press.[3] If New South Wales had possessed a sufficient supply of competent surveyors, clear-headed draftsmen and sensible politicians, his pleading might have produced some effect; but, as things were, it was a waste of his time. Whether he liked it or not, he had no chance of keeping his grass unless he fought for it. He fought for it shrewdly and at times ruthlessly.

Except for 90 acres of freehold around the homestead, Myalla was wide open to free selection when Pratt bought the leasehold. People in Cooma said that he had been a fool not to have made selections on his own behalf before he put his money down.[4] Possibly they were right, but, if they were, he lost no time in learning wisdom. Within the first two months of his precarious proprietorship he secured the freehold of 1,000 acres.[5] He maintained the same cracking pace of acquisition throughout the second half of 1872 and the years immediately following. By and large, his methods were the same as James Litchfield's: sometimes he made a selection in his own name; sometimes he made it in the name of a son or daughter; sometimes he arranged for an employee to make it as his dummy. Not that he ever used that opprobrious word when referring to the Owers brothers, or old Ronald McDonald and his two sons, or any other Myalla man who played straight with him; these men, in his eyes, were good employees. 'Good employees,' he wrote in his diary, 'select for me, bad employees select against me.'[6]

[1] See Dan Coward, 'Free Selecting on the Eumeralla Shore', *JRAHS*, December 1969, table 3 on p. 374. Mr Coward's full and precise account of the battle for grass on Myalla saves me much space. He has, besides, drawn all my maps.
[2] 14 June 1872.
[3] *S.M.H.* 20 April and 4 May 1875. He signed these letters 'A Southern Squatter'.
[4] 14 June 1872.   [5] 6 May 1872.   [6] 20 May 1876.

The amending Act of 1875, with its prohibitions against dummying and selection by minors, opened a new phase – for Pratt, if not for everybody.[1] Thereafter, he secured most of the land he still needed by improvement purchases, or purchases at auction, or by selections in the name of his sons as they reached the stipulated age of sixteen. In this second phase, as in the first, he conducted his campaign from headquarters in Sydney.[2] That proved no disadvantage to him, because the over-centralised Lands Department was the source of all essential information and Pratt was *persona grata* there. He sifted the information quickly, decided his course of action and telegraphed instructions to his brother Sam, who was holding the fort for him on Myalla. In his diary he made a note of every telegram and letter exchanged between Sam and himself: consequently, the diary contains a blow-by-blow account of almost every battle for almost every acre of Myalla.

By 1884 the main results of the fighting were plain to see. An amending Act of that year made provision at long last for a division of every pastoral lease into two zones, one for the lessee, the other for free selectors. But the Act came too late to serve that purpose.[3] In the Cooma district, at any rate, the pastoral leasehold still surviving was barely 4 per cent of the original area. We should therefore be wasting our time if we tried to find any pattern in the few small rectangles marked c on our map of Myalla. We can, however, discover some patterns in the division of freehold between Pratt and his challengers. With this object in view let us take a quick look at map 10.[4]

Starting on the western flank we see at once that Pratt has yielded little ground, except to a selector named Coffey. This man had arrived in the 1840s from famine-stricken Ireland and we may feel sure that he had no wish to meet gentry in his new country. He disliked Pratt's manners; Pratt disliked his. They fought each other with venom. Coffey, to begin with, got rather the better of the fight and built up a freehold of 2,000 acres. What he wanted, however, was not a farm but a pub. So he sold his land and built the Union Hotel in Cooma. Three generations later, his descendants were still in possession there.[5]

Moving along the southern edge of the map we see the marks both of hard fighting and prudent retreat. In the Bobundara valley Pratt has yielded ground to Kiss; but he has fought Smith on the well watered slopes of Rocky Range. Smith managed to accumulate nearly 1,000 acres there, but Pratt blocked his further progress by dummy selections. So

---

[1] 26 May 1881. Pratt notes that dummying continues unabated at Billylingera and Bredbo.
[2] He did not give up teaching and start residing full time on Myalla until 1880.
[3] The Act served two other purposes: (1) it decentralised administration; (2) it established three territorial divisions – Eastern, Central and Western – of the land of New South Wales.
[4] In the article cited on p. 98, n. 1, above, Mr Coward takes a much closer look at it.
[5] Perkins, 2254. Coffey's hotel was later called the Railway Hotel.

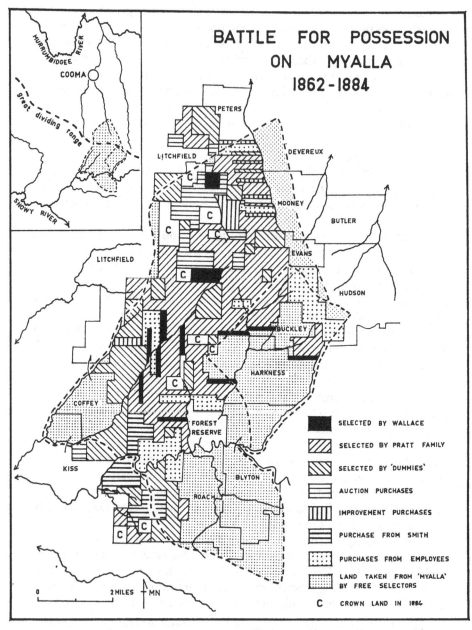

BATTLE FOR POSSESSION
ON MYALLA
1862–1884

SELECTED BY WALLACE

SELECTED BY PRATT FAMILY

SELECTED BY 'DUMMIES'

AUCTION PURCHASES

IMPROVEMENT PURCHASES

PURCHASE FROM SMITH

PURCHASES FROM EMPLOYEES

LAND TAKEN FROM 'MYALLA'
BY FREE SELECTORS

C   CROWN LAND IN 1884

Map 10

Smith, like Coffey, began to think that life in a pub would suit him better than life on the land. In 1878 he sold out to Sam Pratt – at a sufficient price, let us hope, to establish himself in the pub. Fighting Smith had cost the Pratts a lot of money. Wisely, they decided to keep the peace with Roach and Blyton, who had taken selections in the south-east corner of Myalla. Both these men, from the Pratt point of view, were 'good selectors'. Roach, a former shepherd, had been a willing dummy in the fight with Smith; he was always available for seasonal work on Myalla; he sent his daughter into service with Mrs Pratt. Blyton, although he had swallowed a sizeable slice of Myalla, felt no inordinate appetite for land; from time to time he went fossicking for gold.[1] Besides, he and his numerous sons dearly loved racehorses. Neighbours such as these were not dangerous to Pratt.

On the eastern flank, in the beautiful Peak country, Pratt had to face serious danger. William Bradley had advertised the Peak run as a separate property;[2] but Pratt, like the Wallaces before him, bought it as an integral part of Myalla. By then, the selector Kiss had bitten off about 900 acres of the Peak's grass; but Kiss appeared to be content with what he had. The situation changed when William Harkness, whom Pratt was employing as his overseer, put down £2,000 to buy the Kiss selections. That was only the opening move; Harkness wanted more. Pratt tried to block him with dummy selections. Harkness retaliated in kind. In Pratt's eyes, Harkness was a traitor and a cad.[3] Pratt was always apt to use hard words about people whose interests conflicted with his interests; but Harkness – this was the sore point – had been his manager and was, besides, a man of his own class. In the event, honours remained fairly even in the Harkness–Pratt battle. There had been times when Pratt came close to surrendering; but in the event, Harkness had to go elsewhere for a good many of his acres.[4]

Against his own will, Pratt retained possession in the northern sector of Myalla. He did all he could to establish his brother Sam there; but Sam did not survive the hard times that set in during the 1880s. Edward Pratt survived them because he had never allowed himself to run dangerously into debt. The imprint of his prudence can be seen on the map of Myalla; in 1884 he was holding in freehold barely half the area of leasehold that he had bought in 1872. Live and let live had been his rule with free selectors, unless and until their struggles for a place in the sun endangered his own place in it. Not only that: he was invariably cautious in his financial

[1] Perkins, 1118.
[2] In the Bradley advertisement of 1866 it was called Milady Peak. The geographical feature is called Hudson's Peak.
[3] 25 February and 24 April 1878.
[4] 2 March 1878, 14 February 1882. Harkness acquired a property of about 8,000 acres and called it Lincluden.

forecasting and quick to shorten sail if ever he saw dirty weather ahead. Many entries in his diary during the early 1880s illustrate his anxious accounting. The Commercial Banking Company had reduced its mortgage rate from 9 to 8 per cent, but he took note that he had been spending a lot of money on buying and improving land and must now start to 'draw in'. He discussed how best to do so with Beazley, the bank's general manager. After much anxious thought, he told Beazley that he proposed to sell between four and five thousand acres of his freehold. On the last day of 1883 – his birthday – he made this entry in his diary: 'I have a reduced property – but it is in many ways a better one and it ought to be enough for my moderate wants, even though the price of wool has been and threatens to continue depressed.'[1]

In the same entry, he recorded the fulfilment of a dream – he and his wife were back again in England. Their homecoming was such a joy that they were of a mind to stay for ever; but for three troubled years they moved about from place to place without settling anywhere. With wool prices down from 12*d.* a pound in 1880 to 8½*d.* in 1886, Pratt could never stop worrying about Myalla. His wife could never stop worrying about her two grown-up sons, now fending for themselves in a hard world. Both parents felt lonely in the land of their upbringing. In the end, they admitted that they had made a mistake when they set out for England in the hope of renewing old friendships.[2] Towards the end of 1886 they booked their return passages. Friends were on the wharf to meet them when their ship berthed in Sydney. On 12 February 1887 they came home to Myalla. 'Found the old place looking much better than I had expected,' Pratt wrote that night in his diary. 'Fruit trees bearing well. After dinner went drafting and saw stud ewes and lambs which pleased me much.'

It has been worth while to follow closely the struggle for possession on one representative property; but it is now time for us to look at the broad outlines of change on the twenty properties of our sample. What follows will be a brief commentary on two illuminating maps. Map 11[3] shows the succession to William Bradley at the peak of free selection; map 12 shows it thirty years later, when 'closer settlement' – a policy which sought the same end by different means – was just only beginning to get under way. The story these two maps tell can be summarised as follows:

[1] 3 January, 8 and 13 March 1881; 17 January and 16 June 1882; 31 December 1883.
[2] 29 July 1866.
[3] In drawing these maps, Mr Coward used the land title records and the parish maps. The picture of land ownership thus constructed is not complete: for example, it does not measure extensions beyond the Bradley boundaries. Even so, its accuracy is sufficient for present purposes.

Number of properties in the Bradley area

Year	Over 5,000 acres	Less than 5,000 acres	Total
1884	7	76	83
1914	14	38	52

If we compare the map of 1884 with that of 1860 (map 9, p. 93) our first impression will be of good progress achieved in redistributing property rights. Only seven of Bradley's twenty stations have survived: of these seven, the majority are much smaller than they had been before. Myalla, as we have already seen, has been cut down to half size: some stations, such as Dangelong, have suffered even more drastic surgery. Room has thus been made for nearly four-score new properties, ranging in size from a couple of hundreds of acres, or less, to a couple of thousands of acres, or more. The free selection policy, it would appear, is making fair progress towards its objectives. But the 1914 map tells a different story: the four-score properties that were on the map of 1884 have been cut to half that number; only one in three of the properties below 5,000 acres has survived; there are now twice as many properties above 5,000 acres. Plainly, the main trend between 1884 and 1914 has been towards the aggregation, not the disaggregation, of property rights.

A closer look at the map of 1884 will show that this trend was operating even then, side by side with the trend towards dis-aggregation. Enough has already been said, at least for the time being, about James Litchfield's acquisitions at the northern end of the Bradley territory.[1] At the southern end, Bibbenluke bestrides the pastoral map like a colossus. For three decades or more, Bibbenluke was the operational base of an extraordinary man, Henry Tollemache Edwards, manager of the Bradley estate in Monaro. The records kept with meticulous care by Edwards make it quite clear that he managed, not only the estate, but also the trustees. In the late 1860s he rubbed out the boundaries of the five Bradley stations immediately to the north of Bibbenluke, thereby creating a single centralised property of 60,000 acres. Throughout the 1870s and early 1880s he converted that property piece by piece into freehold land. When the trustees took fright at the immense financial cost of this operation, he bullied or wheedled them into giving him a free hand. With equal success, he bullied or wheedled his dummy selectors. 'The way in which I secured the run here,' he wrote on 20 August 1889, 'by the management of the Selections without one case of a man playing me false, is still a wonder to the whole country!' It ceases to be a wonder when one studies Edwards' carrot-and-stick method of handling men; few of the dummy selectors would feel a strong temptation, none of them would feel brave enough, to play him false. While he was making Bibbenluke secure, he was doing the same thing nearby at Burnima, a large property which his mother's second

[1] See pp. 92–6, above.

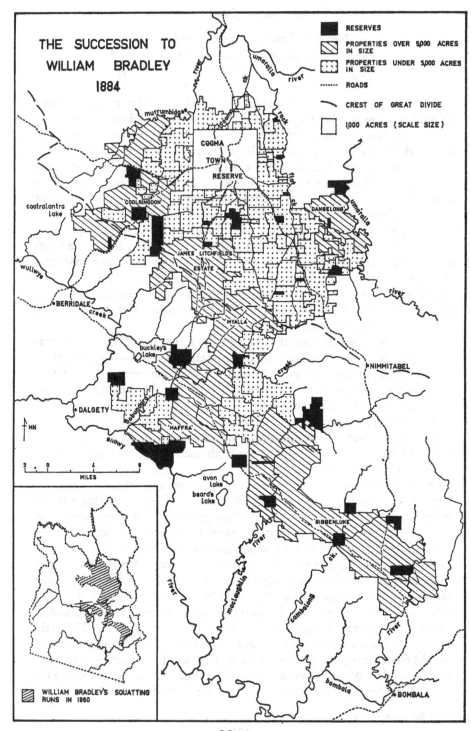

THE SUCCESSION TO
WILLIAM BRADLEY
1884

RESERVES

PROPERTIES OVER 5,000 ACRES
IN SIZE

PROPERTIES UNDER 5,000 ACRES
IN SIZE

ROADS

CREST OF GREAT DIVIDE

1,000 ACRES (SCALE SIZE)

murrumbidgee river

umaralla river

COOMA
TOWN
RESERVE

cootralantra
lake

'COOLRINGDON'

'DANGELONG'

umaralla

JAMES LITCHFIELD'S
ESTATE

wullwye

river

BERRIDALE creek

MYALLA

buckley's
lake

NIMMITABEL

DALGETY

creek

MN

MAFFRA

snowy

BIBBENLUKE

2   0   4   8
MILES

avon
lake
beard's
lake

maclaughlin river

cambalong ck.

river

river

bombala

BOMBALA

WILLIAM BRADLEY'S SQUATTING
RUNS IN 1860

Map 11

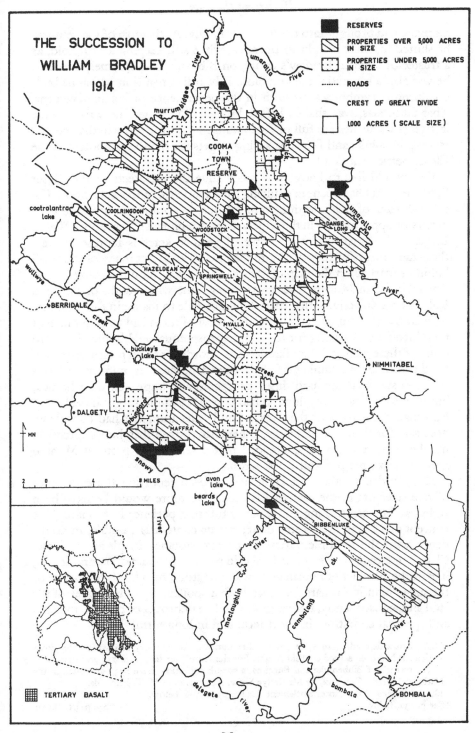

THE SUCCESSION TO
WILLIAM BRADLEY
1914

RESERVES
PROPERTIES OVER 5,000 ACRES IN SIZE
PROPERTIES UNDER 5,000 ACRES IN SIZE
ROADS
CREST OF GREAT DIVIDE
1,000 ACRES (SCALE SIZE)

murrumbidgee river
umaralla river
rock ck.
flat ck.

COOMA TOWN RESERVE

cootralantra lake
'COOLRINGDON'
'WOODSTOCK'
DANGE-LONG
umaralla river

wullwye
'HAZELDEAN'
'SPRINGWELL'

BERRIDALE
creek
'MYALLA'
NIMMITABEL

buckley's lake
creek

MN
DALGETY
'MAFFRA'

2 0 4 8 MILES

snowy

avon lake
beards lake

river

BIBBENLUKE

maclaughlin river
cambalong ck.
river

TERTIARY BASALT

delegate river
bombala river
BOMBALA

Map 12

marriage had brought into the Edwards family. At the turn of the century, he shifted his main weight to Burnima, while still keeping a firm hold, as a trustee of William Bradley's estate, on Bibbenluke. Supremely a realist, he was also a romantic: the grandiose Edwardian mansion which he built on Burnima symbolises his faith in the permanence of his achievement. That faith, alas, was shaken before he died. The letters he wrote in the last years of his life are full of foreboding. He learnt too late that he had made Bibbenluke and Burnima sitting targets for the sharpshooters of the Closer Settlement Act.[1]

One would like to know how much time the architects of the Closer Settlement Act had spent in studying the experience of free selection.[2] The achievements of free selection were various, in accordance with the wide variety of opportunities in the widely diverse regions of New South Wales. In pastoral Monaro – if our sample of 300,000 acres is representative – the achievements were never more than precarious: not many of the small holdings marked on the map of 1884 can still be seen on the map of 1914. The majority of Monaro's selectors never came within sight of Sir John Robertson's objective – that they should be able to maintain their families by their own labour on their own land: most of them had to earn much of their income by working for their more prosperous neighbours. The hard times of these men and their families are recorded, not on parish maps, but by the rubble and lumber of their ruined homes.[3] A minority of free selectors survived the hard times, prospered and improved their homes; but they would not have done so had they not acquired a good deal more land than the makers of the Crown Land Alienation Act had considered necessary for them. By and large, experience proved that James Litchfield had been right when he contended that a selector in pastoral Monaro needed between 2,000 and 3,000 acres.

He would not have been right had he been reporting the experience of free selectors in the eastern Riverina.[4] No more would he have been right had he been making a forecast of future experience in Monaro. The concept of a living area seems clear; but its content is variable, in accordance with the multiple variations of circumstance in place and time. Changing circumstance on the land is in part the product of natural forces, in part the product of human action. In our story, man holds the centre of the stage. Man in Monaro has been both a spoiler and an improver of the land and, consequently, both a spoiler and an improver of his own material and spiritual condition, in short term and in long term.

[1] H. T. Edwards died in 1915, five years after notice of compulsory acquisition for closer settlement had been served on Bibbenluke. For the sequel, see below, p. 115 and map 14.
The records of Bibbenluke and Burnima, a superb collection, have been deposited in the National Library, Canberra, by Mr Battye Shaw, a grandson of H. T. Edwards.
[2] On the origins of the Closer Settlement Act see pp. 155–6, below.
[3] See p. 132, below. [4] See p. 91, above.

2. Spoiling

Westwards and southwards from William Bradley's territory the Snowy River makes a majestic bend to the east, the south, the west, until it settles at last on a steady course to the Southern Ocean. Within this bend the land is high, open and, in good seasons, well grassed. Thither, in 1835, Amos Crisp and his three brothers drove their cattle until they came to Jimenbuen. After a few years the younger brothers went their several ways, to the west and south, leaving Amos in sole possession of Jimenbuen, Matong and Numbla. Together, these three properties aggregated at the very least 70,000 acres.[1]

Amos Crisp married and produced twelve children. The seventh, William, possessed a photographic memory and the story-teller's gift.

> Jimenbuan in the early days was very different from what it was after the passing of the Sir John Robertson's Land Act, which gave selectors the privilege of taking the land selected before survey. Some of them would put on more stock than the area they selected would carry . . .
>
> Before the passing of the Land Act . . . the Matong Creek for about five miles above and below its junction with the Jimenbuan Creek was a succession of deep waterholes, there being no high banks, and grass grew to the water's edge. Hundreds of wild ducks could be seen along these waterholes, and platypus and divers were plentiful. Five years after the passing of the Act the whole length, instead of being a line of deep waterholes, became a bed of sand, owing to soil erosion caused by sheep. The water only came to the surface in flood time, when it spread the sand over the flats. That was the actual state of affairs when I left Jimenbuan in 1899, and I have no doubt it is the same today.
>
> . . . Free selection and the introduction of sheep put an end to extensive raising of cattle because of the loss of water in the main creek and because the best grasses for cattle disappeared and made way for a finer and, in my opinion, less nourishing grass. I have seen the kangaroo grass, when in seed, like a field of wheat three feet high. This disappeared, as also did many of the wild birds, including the duck and the turkey.[2] I well remember the old wild turkey cock strutting around with his white crest let down. Then came the starling, and later the hare and the rabbit. The curlew disappeared

[1] *Early History and Incidents in the Life of William Crisp* (M.L. typescript signed by William Crisp and dated 25 July 1947). The Aborigines said that Jimenbuen means 'bit fat kangaroo rat'. The usual spelling of the name is with an *e*, not with an *a*, as Crisp spelt it.
[2] I.e. the bustard.

with the coming of the fox. Then came the fruit pests such as the codling moth, the fruit fly, the silver-eye and the bulbul. These bring to mind the old orchard with its apples, peaches, cherries and plums, for which there was no market. We boys used to shake a peachtree and bring down a dozen or so of the ripe fruit; we would take one or two and leave the rest for the young pigs.

We need not take as gospel truth everything that William Crisp tells us in his lament for a vanished Eden. He was an unsophisticated observer, as we may see from his list of bad birds, in which the delicate Australian silver-eye keeps company with that coarse immigrant, the starling. He was also a prejudiced observer, a cattleman who hated sheep, a squatter who abhorred Sir John Robertson and his Land Act. We must heavily discount these prejudices. Cattle, we may agree, had been good pioneering stock; they could travel long distances to water; they were safe, or nearly safe, from the dingos; they could fend for themselves in rough country; they made it easy for the squatter to cut his labour costs. Even so, William Crisp appears to be excessively indulgent to cattle: they, not the sheep that he so disliked, were the main destroyers of his beloved kangaroo grass. In destroying it, they made the pastures more hospitable to sheep and thereby enlarged the grazier's freedom to make a profitable choice between the opportunities of the market.

We must still more discount William Crisp's prejudice against free selectors. He seems to take it for granted that all was well with the land until they descended on it; but we have seen already that the opposite was true. On the eve of Sir John Robertson's land legislation, an official investigator had diagnosed 'the malady of mismanagement' as the deadliest disease afflicting cattle and sheep.[1] Twenty years later, the Commissioner of Crown Lands in Monaro considered squatters and selectors equally at fault in spoiling the pastures by overstocking them.

> There is a very great increase in the number of stock now kept on the same area of land, than formerly, throughout the whole of Monaro, on those parts where selection has been going on rapidly, as the lessees of the runs have not decreased their stock in the same proportion that large areas of their runs have been taken from them by conditional purchasers, and as most of the selectors have got sheep the land has been made to carry twice the number it formerly did, and which overstocking is, and has done, an immense harm to the grazing capacities of the country, and from which cause I consider may be attributed the great losses that occured in the dry season of 1878.[2]

The Commissioner, H. K. M. Cooke, writes curious English; but his message comes through. If we look for guilty men, we shall be wasting

[1] See above, pt. II, 2, esp. p. 69.
[2] *Occupation of Crown Lands, Annual Report, 1879* (Sydney, 1880), pp. 150–3. This report was the first of a series, but was not printed with the Parliamentary Papers because it was produced before the Resolution of the House of Assembly (11 November 1879) which stipulated tabling and publication.

our time. Later on we shall look for ignorant men and shall find them, not only among the squatters, selectors and politicians, but also in the academic fraternity. But let us stay content, for the time being, to look for men under economic pressure. Squatters and selectors were under the same pressure to find money for land purchase and for the improvements which the law required. Consequently, they were under the same pressure to squeeze from the land every pound of immediate income which it would yield. To resist these pressures required quite exceptional intelligence and self-restraint.

Notwithstanding his prejudices, William Crisp was an accurate and vivid reporter of things that he had seen with his own eyes on his own land. His report of change on Jimenbuen suggests the agenda for two important inquiries into man's impact upon the natural environment during the half century which followed the Crown Lands Alienation Act. In the first inquiry, our concern will be soil cover, soil moisture and the flow of water; in the second, it will be animal life.

Let us start the first inquiry at Matong Creek. Formerly it had been a chain of ponds, with a quiet glide of water from pond to pond and grass growing to the water's edge. Trampling by stock damaged the grass, laid bare the soil and changed the quiet stream into a deeply incised water-course, bone dry except after heavy rain, when the rush of water scoured the banks and braided the bed of the creek. Not far away, on Boloco station, similar causes produced similar effects.[1] To this very day, a traveller along the road from Canberra to Cooma will see a comparable degradation of almost every watercourse that he crosses. At Billylingara near Bredo, notwithstanding some patient repair work during the 1950s and 1960s, the scoured gulches and eroded hillsides are horrible to behold; on Carlaminda station, upstream from the Umaralla Bridge, we see a river bed that looks like a battlefield.

It would be difficult to establish the chronology of afflictions such as these, because their origins and onset were seldom noted by contemporaries. To this statement we may make one exception: a good many people observed the evil effects of overstocking. From the late 1860s to the early 1900s they were recorded, almost as a matter of routine, in the annual reports of Alexander Bruce, the Chief Inspector of Stock. Year by year he lamented the destruction of good pasture and the multiplication of inferior plants and noxious weeds, both native and exotic. Unfortunately, he seldom cited evidence from Monaro: still more unfortunately, the annual

---

[1] Written information from Mr H. A. Rose. This damage to Boloco Creek was done in the 1880s – by cattle, not sheep. In other places, and at a later time, rabbits did most of the damage. In other places still, erosion started with the plough or with the tracks made by carts: cf. Samuel Schumack's *Autobiography* (ed. L. F. Fitzhardinge, A.N.U. Press, 1967), pp. 108–10, 208.

reports of his assistant there have been lost or destroyed.[1] We must therefore go fossicking for such bits and pieces of information as have been preserved in the press, in diaries, or in station records. Sometimes they enlighten our ignorance. We know, for example, that James Litchfield was putting men to work during the 1880s on the eradication of Bathurst burrs and star thistles, and that his land became heavily infested, then or a little later, by that nasty native pest, corkscrew grass. Its ravages were prominent among the pressures which impelled him every summer to drive large flocks of sheep to the Alpine pastures.[2]

It would have been surprising if practical men had failed to observe and lament the spoiling of their pasture. It would have been just as much surprising if they had studied with close attention the soil beneath the pasture and the water in the soil. William Crisp observed a conspicuous example of soil erosion but had no idea at all of its scientific explanation. There was nobody to teach him. Up to the last decade of the nineteenth century, soil science and hydrography held no place either in the educational curriculum or in the public service of New South Wales.[3] The colony did, however, possess some broadly educated and highly gifted men of science. We have already made the acquaintance of A. G. Hamilton and shall now consider further the very informative paper which he read in September 1892 to the Royal Society of New South Wales.[4] Among many other interrelated phenomena, he considered the effects produced by men and their animals upon the soil.

> When the surface is broken on a slope, no matter how gentle, the protection afforded by the grasses and herbaceous plants to the soil is removed and the surface drainage is altered. Small runlets of water begin to travel along the line of disturbance and to cut channels which become deeper and deeper. The amount of earth cut away of course depends greatly on the slope, the nature of the soil and the amount of rainfall, being greatest in light soils and on steep slopes ... When steep hillsides have been cleared, the roots left in the ground decay, and the binding of the particles of mould which they effected is lost; when a wet season comes, the upper soil resting on rock or on a stiff sub-soil, becomes saturated with water till it is in a viscous state and moves down the hillside slowly; an unusually heavy downpour of rain when it is in this condition, precipitates landslips down the hillsides, and considerable areas slide down, overwhelming the herbage which grew on them and that on the level ground at the bottom in one common destruction.

[1] I am indebted to the New South Wales Archives for making a pertinacious search for these reports; but none from any pastoral district have survived.
[2] Information from Mr J. F. Litchfield of Hazeldean. Corkscrew grass (*Stipa setacea* R.Br.) was unnutritious and produced a fine seed which not only got into the wool, depreciating its value, but often got into the eyes of lambs, blinding them and even killing them. The seed of barley grass and eight-day grass also caused losses, albeit less drastic ones.
[3] See pp. 119–20, below.
[4] See pp. 62–3, above.

Later in the paper, Hamilton made a rough balance sheet of the nutrient substances which cattle and sheep subtracted from the soil and those which they added to it. As before, his conclusions were sombre.[1]

Contrariwise, the conclusions of W. W. Froggatt, President of the Linnaean Society of New South Wales, were sanguine. He said in effect, as farmers used to say in East Anglia, that the grazing animal has a golden hoof.

> The next important change brought about in new country by stocking, was the hardening of the soil, the eating off of the rough grass, and the consequent improvement of the pasturage. This is known to every stockman, and I had a striking example brought under my notice on the King's Sound Pastoral Company's station in North-West Australia, where, in the midst of about two million acres of unstocked land, there was a fenced-in paddock of about ten thousand acres, upon which the station sheep were depastured. Within three years, the enclosed land, though very slightly stocked, was transformed into a different and better class of country from that outside the ring-fence, which was poor and thinly grassed.

To put livestock on the land, Froggatt continued, not only made the pasture better but made surface water run where it had never run before.[2]

What are we to make of such conflicting testimony? Possibly we may believe both witnesses, for their diverse conclusions were based on observations made in widely diverse environments. The observations of Hamilton are more closely relevant than those of Froggatt to conditions in some parts of Monaro; even that small province contains many contrasting environments. For example, the Umaralla in its middle reaches is a wreck; but its main tributary, the Kydra, is an unspoilt stream. How has that contrast come about? Since the pressure of human and animal population on the land was just as heavy upstream as it was downstream,[3] it would appear that the land upstream was less vulnerable to pressure. Superficial observation supports this hypothesis – the Kydra valley has the higher rainfall, the firmer soil, the tougher root system and the more plentiful supply of water weed. Superficial observation is not, however, good enough, either for the ecologist or the historian; neither will get very far in his search for knowledge except by the patient and precise exploration of small areas. Sometimes the exploration may yield an extra profit if it is planned and conducted as a combined operation.[4]

[1] *On the Effect Which Settlement...*, pp. 189–90, 213–16.
[2] *Proc. Linn. Soc. NSW* xxxviii (1913), 'A Century of Civilisation from a Zoologist's Point of View'. See particularly pp. 22–4.
[3] Free selection destroyed the Kydra run: twelve freehold properties were cut out of it and only three of the twelve were larger than 1,000 acres. Nearly all the selectors had very large families; all of them had cattle, or sheep, or both.
[4] See Appendix 2, Tasks (pp. 200–2, below).

Possessing the land

Today, nearly all the small homesteads of the Umaralla and Kydra valleys are mouldering into the ground. Not the slump, nor the drought, but the rabbits, beat the battlers on those selections. So we have to find out when and how the rabbits invaded Monaro. This inquiry cannot be rushed: we have first to find out how the native animals had been faring. William Crisp's account of ecological change on Jimenbuen is our point of departure; but we know from previous experience that it will have some blind spots. In particular, Crisp tell us nothing at all about that most conspicuous ecological disclimax and human tragedy, the destruction of the native men who had hunted the native animals.

All the white people of Monaro had the same blind spot. We could read tens of thousands of pages of newspaper print without tracking down a dozen references to the last agonies of the black people. On 27 September 1856 the Monaro correspondent of the *Sydney Morning Herald* reported their 'almost entire extinction': the few survivors that were sometimes seen belonged to the Omeo or Gippsland tribes: they did some desultory stock work for a trifling remuneration: they had an inordinate love of strong drink and no belief in God. Journalism such as this tells us more about the whites than about the blacks. The latter were not yet so close to extinction as the Monaro correspondent reported: according to the census of 1856 there were 166 Aborigines in the Cooma district and 319 in the Bombala district. Thereafter, their decline was rapid. On 4 July 1867 the *Monaro Mercury* published the following snippet of news: 'Several aborigines, remnants of a once numerous tribe, have been promenading the town, and on Monday last were the recipients of the usual donation [a blanket] contributed by the government to the natives of this colony.' In the early 1880s the government established an Aborigines Protection Board. It possessed no local staff but used the police to distribute the government's bounty and to collect vital statistics. According to the returns thus obtained, the entire Aboriginal population of New South Wales in 1886 was 7,634; to this total, the contribution of Monaro was zero. But the police constable of Cooma had done his counting carelessly. There still survived two members of the Ngarigo tribe, Bony Jack and his son Biggenhook. From an occasional paragraph in the press we get the impression that the white people made Bony Jack a figure of fun, but felt some affection for Biggenhook. He was deaf and dumb; but he was an intelligent man, a willing worker and an ardent fan of Cooma's cricket and football teams. In June 1914 he died, and with him died the Ngarigo.[1]

The extinction of the first discoverers of Monaro seemed to their white successors a natural occurrence; but one of its side-effects troubled them.

[1] Perkins, 957, 978, 1084, 2597, 2598. V.P. Leg. Ass., 1884, vol. 3, p. 900; 1885, vol. 2, p. 605; 1887, vol. 2, p. 855.

Spoiling

As the Aborigines dwindled, the marsupials returned in a flood – in 'plague proportions', the pastoralists said.[1] Men of science used the same phrase and found the same explanation of this unexpected and unwelcome natural phenomenon. To quote Froggatt:

> We can now consider the most striking changes that came with civilisation, and the passing of the aboriginals as a nation of hunters, the first of which was the enormous increase of the indigenous animals and large birds, not only caused by the disappearance of the native, but also by the partial extermination of the dingos and wild dogs, both of which lived upon the native fauna . . . This remarkable increase of marsupials, in particular, was very noticeable even in the early fifties . . . The opossum, like the kangaroo, lost its enemies, and multiplied rapidly in all suitable localities.[2]

To protect their grass, crops and gardens, the owners organised hunting drives and spread poisoned baits. Parliament took a hand in the struggle, starting, as one would expect, with legislation against the dingos – for they killed sheep – and continuing with measures against the marsupials and other 'noxious animals'. In successive Acts and amending Acts throughout the second half of the nineteenth century a code of practice took shape: first, responsibility was placed on the landowner to keep his land 'clean' at his own expense; secondly, inspectors were appointed to see that he did so; thirdly, powers were vested in elective boards of directors in the pastoral districts to supervise operations and to impose on all landowners a levy that would cover part at least of the costs. As operations proceeded from decade to decade the government found itself increasingly constrained to supplement these local efforts by a direct subsidy.[3]

Most conspicuous among the items of cost was the blood-money paid to persons who produced the scalps or paws or tails or talons of animals that had been declared noxious. Sceptics sometimes pointed out that the bonuses gave wily individuals a motive for maintaining or even fostering the populations of noxious animals. Be this as it may, the annual reports of the Chief Inspector of Stock from the 1880s to the 1900s contain astonishing tallies of slaughter – in most years, kangaroos and wallabies by the million; kangaroo rats in hundreds of thousands; opossums, bandicoots, paddymelons and crows in scores of thousands; eagles, hawks and emus

[1] E.g. Samuel Schumack's *Autobiography*, p. 152, recalling the 1870s and 1880s: 'I first saw a kangaroo in 1867, and now they were in plague proportions.'

[2] Froggatt, 'A Century of Civilisation from a Zoologist's Point of View', pp. 18–21. Man-induced changes of the pasture should be added to his recital of causes.

[3] The main legislative landmarks are the Native Dogs Act 1852, the Pastures and Flock Protection Act 1880 (this Act included rabbits for the first time) and the Pastures Protection Act 1902. See *Manual of the Pastures Protection and Stock Acts of New South Wales* by W. S. Acocks and J. E. Clark (Sydney, 1907).

in thousands; wombats in hundreds. To have left the wombats off the list would have looked careless.[1]

The killing was long-sustained and indiscriminate. We need not doubt that it achieved its immediate objective of saving crops and pasture for the use of men and their domesticated animals. On the other hand, no attempt was made at that time to count the cost in terms of the damage done to Australia's unique fauna. What is more surprising, only a few people attempted to count the cost in terms of utility. Along this small minority, as we should expect, was that sophisticated Darwinian, A. G. Hamilton, who explored the possibilities of a relationship existing between the declining numbers of insect-eating animals and the increasing incidence of plant diseases. This novel habit of observation and reasoning began at long last to find occasional expression in the columns of the *Pastoralists' Review*. One correspondent suggested that the best way of coping with liver fluke would be to protect the wading birds that ate the snails that were hosts to the fluke embryos. Another observed that there had never been grasshopper plagues so long as the native turkeys survived to gorge themselves on the wingless young of the swarms. 'We have the rabbit plague,' the same writer declared, 'because we have first exterminated by poison all their natural enemies.' Possibly he over-rated the death-dealing capacities of the eagles, hawks and marsupial cats; yet it does seem strange that the man on the land so seldom saw those creatures as his natural allies in the ruinous rabbit war. On Myalla, Edward Pratt waged implacable war against marsupial cats. In the *Pastoralists' Review*, Mr T. Shaw proved by figures that strychnine was the proper medicine for eagles: 'Any man can poison; every man cannot shoot straight. For us the easiest way to get rid of eagles is poison.'[2]

The last sentence epitomises the majority view at the turn of the century: consequently, the crudities of an earlier time need not surprise us. In the *Annual Report of the Government Botanist of Victoria, 1861–62*, we read that selected blackberry seed has been sent to interested recipients throughout the length and breadth of the land.[3] The author of the report, Baron Sir Ferdinand von Mueller, F.R.S., did not foresee how little time it would take for the blackberries to make a break-away in well-watered districts and for the rabbits to find safe shelter under their thorns. We have no just cause to think ourselves wiser or more virtuous than von Mueller:

[1] I had it in mind to graph the annually reported killings, under the main heads, for the forty years 1882–1911; but decided that my graphs would be misleading: first, because the bonus payments often understate or overstate the numbers of animals killed; secondly, because the figures have reference not to Monaro but to the whole of New South Wales.

[2] *Past. Rev.*, 15 October 1891, 15 June and 15 August 1896, 15 June 1899. Pratt, 15 May 1882 and 12 February 1887.

[3] P. 6. A good account of von Mueller's life and work, by Charles Daley, will be found in *Vic. Hist. Mag.* x (1924).

his journeys of scientific exploration were heroic; his name still shines in the annals of Australian botany; if he made any other ecological errors, they are not on record. The same cannot be said for von Mueller's contemporary, Frederick McCoy, Melbourne's first professor of natural science. He, too, achieved academic recognition and a knighthood; but he was an ass. He produced learned works on Australian palaeontology and zoology but walked blind and deaf through the Australian Bush. It was cursed, he declared, by a savage silence; but the day was dawning when acclimatised birds, 'delightful reminders of our English home', would civilise it. Some of the birds on his list would minister to the spiritual needs of Australians by filling the air with 'Heaven-sent Melody'; others would kill grubs. He felt deeply hurt when a fellow-member of the Victorian Acclimatisation Society complained that the proliferating sparrows killed no grubs but plundered many orchards.[1]

The word acclimatisation has various connotations. It is, to begin with, a useful word for summarising a process which has been continuous from the dawn of history up to the present – the migration of men, plants and animals to new environments. There is no region of the earth (except, perhaps, the polar regions) which does not now possess large numbers of acclimatised species. Some migrations have been involuntary: one recalls a story of the Roman soldier who made the passage to Britain with Gallic soil sticking to his boot – 'next year, Britain had another wildflower'. Other migrations have been planned; corn and cattle did not come to Britain, nor did dingos come to Australia, without deliberate acts of human will. Governor Phillip brought with him to Australia many and various plants and animals, including rabbits. Acclimatisation, it thus appears, is not only a process but a programme; Sir Joseph Banks said that it was his business to encourage the transport of plants from one country to another. This business is useful when it is prudently pursued; but Europeans of the mid-nineteenth century pursued it recklessly. They inherited a programme and made it a craze.

In 1854 enthusiastic Frenchmen established *La Société Impériale d'Acclimatisation*. British enthusiasts did not follow suit until the early 1860s, but in the meantime found other means of promoting the good cause. In January 1860 Professor Owen, a zoologist of London University, feasted the friends of acclimatisation on roast eland. In a letter to *The Times*, he described the dinner as a gastronomic triumph and appended a preliminary list of the animals, birds and fish which England needed for her

---

[1] *The Australasian*, 20 May 1899, obituary of Sir Frederick McCoy. Cf. Eric C. Rolls, *They All Ran Wild* (Angus and Robertson, 1969), pp. 223, 229–30, 232, 237–9, 244–5, 272. This book is worth reading from cover to cover. It contains far and away the best account, so far, of the rabbit plague in Australia. No single individual bears responsibility for that disaster; Thomas Austin's most conspicuous folly was self-advertisement. Many other people had tried to acclimatise rabbits; Austin was not the only man who succeeded.

embellishment, commerce, sport and gastronomy. To an Australian participant at the feast the list appeared conspicuously at fault in its forgetfulness of middle class families; the sheep was too large, the rabbit too small for their dinner tables; what they really needed was the wombat.[1] The Australian was Edward Wilson, editor of *The Argus* and an influential citizen of Melbourne. On his return from London he founded the Acclimatisation Society of Victoria. Within the next few years sister societies were zealously at work in New South Wales, Queensland and South Australia. They brought to Australia some beautiful and useful animals and plants. They also brought rabbits, racoons, mongooses, agoutis, dandelions, blackberries and the prickly pear. Some of these importations failed to thrive; others throve too well.[2]

We may seem to have been roaming too far from Monaro; but ideas conceived in distant places made history there – or might have made it. Among the might-have-been are llamas. In January 1860 the Monaro correspondent of the *Sydney Morning Herald* announced two items of glorious news: gold had been found again on the Snowy Plains: the llamas were on their way to Nimmitabel.[3] They were under the care of Charles Ledger, Superintendent of Llamas in the colony of New South Wales. Two years previously Ledger had landed in Sydney with a mixed flock, about 300 strong, of llamas, alpacas and vicunas.[4] His alpacas made obvious sense, for their fine wool was in strong demand among English manufacturers of light luxury suitings; but his llamas made dubious sense. The Peruvians used them as beasts of burden, wove their coarse wool, ate their flesh and burnt their dung as fuel; but New South Wales already possessed a sufficiency of fuel, meat, coarse wool and draft animals. Ledger, however, had it in mind to produce a cross which would combine and magnify the separate virtues of the different breeds. According to his reports, he was making good progress with his genetic experiment; but meanwhile the flock dwindled. The southern tableland of New South Wales, we may assume, was ten thousand feet too low for the comfort and health of llamas. They never got so far as Nimmitabel, but got stuck at Marulan. After a year or two, the government sold the beasts at a heavy loss and Ledger returned with shattered fortunes to South America. His reverse in New South Wales did not dismay him. He had ahead of him a dazzling

[1] F. Buckland, M.A., student of Christ Church, Oxford, reported these stirring events in a paper read before the Society of Arts in November 1860. Next year the paper was reprinted by the Victorian Acclimatisation Society.

[2] The Victorian Acclimatisation Society played no part in bringing rabbits in, but McCoy and Wilson expressed satisfaction at their successful acclimatisation. The Queensland Society imported both rabbits and hares – and, of course, the prickly pear.

[3] Perkins, 569, 595.

[4] The flock included very few vicunas and no representatives at all of the closely related guanacos.

triumph – the acclimatisation of quinine in Java. *Cinchona ledgeriana* commemorates an English hero who served humanity well.[1]

So Monaro people never got llamas – which they wanted. Instead they
got rabbits – which they did not want. Nevertheless, they enjoyed quite
a long period of immunity from the plague. The rabbits made their main
advance northwards across the red basalt plains of the Western Division
of New South Wales, with subsidiary advances from small patches of
infestation around Sydney and the lower reaches of the Shoalhaven River.
A map which the Chief Inspector of Stock prepared in 1882 shows them
already well-established in the Riverina, but still a long way from Monaro.[2]
We can imagine Monaro people saying to each other – 'It can never happen
here.' Actually, there were signs during the 1880s that it was already
beginning to happen; rabbits were seen from time to time in the neighbourhoods of Adaminaby, Cabramurra and Cooma.[3] These isloated
colonies did not, however, make a break-away; perhaps the eagles, hawks
and marsupial cats still remained numerous enough to cope with them. So
the pastoralists of Monaro felt safe and felt themselves unfairly put upon
when the government refused to exempt them from the obligations imposed by the Rabbit Nuisance Act. Their *clean* district, they said in effect,
should not be asked to share the burdens of *dirty* districts.[4]

During the 1890s hares, not rabbits, were proliferating in Monaro.
Shooting drives on Coolringdon, Hazeldean, Jimenbuen and other properties achieved bags of 300 and more; in 1897, 5,000 hares were sent by
train to the Sydney market. In comparison with the later tallies of frozen
rabbits, that figure appears trivial. Hares at worst were a nuisance, not a
serious danger; they attacked the cultivation paddock and the lucerne field
but did not despoil the pasture. As A. G. Hamilton put it, 'Pastoralists
complain of the rabbits, farmers of the hares.'[5] Monaro had few farmers.

In the early years of the new century the rabbits began, at long last, to

---

[1] See particularly V.P. Leg. Ass., 1859–60, vol. 4, pp. 997–1005, *Correspondence Respecting
Alpacas*, and Phyllis Mander Jones, 'A Sketch Book Found in Australia' in *Inter-American
Review of Bibliography* III, no. 3, (1953) 280–8. Scores of pamphlets and papers on alpacas were
published in the mid-nineteenth century. Thesis writers, keep off the grass! We want to meet
Charles Ledger in imaginative drama. The dramatist would have to spend some time on the
roof of the Andes and make himself acquainted with its people and animals.
[2] V.P. Leg. Ass., 1882, vol. 4, pp. 1509 ff. Alexander Bruce submitted this map with an important and urgent report which led to the Rabbit Nuisance Act of 1883. There was an
absurd sequel to the Act. Under section 31 the Governor issued the following proclamation:
'Now, therefore, I, the Right Honourable Lord Augustus William Spencer Loftus, with the
advice of the Executive Council, do hereby declare the domesticated cat to be a natural
enemy of the rabbit, and prohibit the killing or capturing of any such animal.'
[3] Perkins, 1634, 1643–4, 2059, 2142.
[4] Edward Pratt (diary, 16 March 1886, 16 September 1856) resented and resisted the rabbit
levy. Cf. V.P. Leg. Ass., 1890, vol. 5, p. 328, a petition by stockowners in the Cooma district
for repeal of the Act.
[5] Perkins, 2025, 2046, 2068, 2078, 2085, 2104, 2169. Hamilton, *On the Effect Which Settlement . . .*,
p. 209.

swarm in Monaro. Broadly, it would be true to say that small men in poor country found the swarms more than they could cope with; but large proprietors in good country, provided they were competent managers, gradually brought them under control. The degrees of disaster or survival were diverse, in correspondence with Monaro's diversity of ecological and economic environments; a realistic estimate of the damage done by the rabbits would require many local studies. All that we can do is to paint the contrast between two extremes, the starveling Kydra district and the rich basaltic plains around Cooma. Selectors had come late to Kydra station because it possessed no really good land, except on the valley floor. In the middle 1880s they were ringbarking the trees to get more grass; but in the slump of the 1890s they allowed the trees to sucker and the seedlings to spring up, until the forest became far thicker than it had been before. From their entrenchments in the forest fringes, the rabbits attacked the patches which the selectors had managed to keep clear. Hitherto, the selectors had eked out a living by summer dairying, the sale of calves to the butchers, possum shooting and seasonal work for the larger land-owners; but the rabbits undercut their livelihood. They could seldom find buyers for their properties; but they could walk off them and in a colony that was starting to expand again they had places to walk to – jobs on the railways, in the police, in the shops of country towns and the factories of Sydney. A few hung on and exploited the rabbits as an industry: the nearby butter factory, instead of closing down, froze rabbit carcases. Thus did the pastoral economy slip back towards the earlier stage of hunting and gathering.[1]

Large landowners on the basaltic plains also suffered heavy losses; but they commanded greater resources and had an easier problem to cope with. Except in the granitic outcrops, the rabbits had no good cover. The landowners enclosed their properties with rabbit-proof wire netting; they sent the poison carts around to lay trails of bran, pollard and phosphorus or to drop a mixture of jam and strychnine on sods that had been turned up; they let their dogs loose on the rabbits and set spring traps. Having thus gained a measure of control, they set themselves to the task of digging the survivors out and destroying all cover. Their struggle for survival was bitter and long drawn out. 'It was not until 1930,' one of them has written, 'that the hunt began for the last rabbit, a hunt which continues to this day.'[2] During the 1950s, myxomatosis did deadly execution among the rabbits; but the improving pastoralist realises today that he must remain forever vigilant.

[1] Information from Mr H. A. Rose, and *Cooma–Monaro Express*, 6 May 1960, an interview with Mat Tracey.
[2] Information from J. F. Litchfield.

3. Improving

In 1870 a young man named William Farrer arrived in New South Wales. He was a farmer's son, a Cambridge graduate and – although neither he nor anybody else knew it – a man of genius. Had his health been good, he would have stayed in England to practice medicine; had he not suffered a financial loss, he would have established himself quickly as a colonial landowner. For Australia, these early mischances of the young man proved to be a stroke of luck.

He found himself attracted to the southern tableland of New South Wales and took a position as tutor in the Campbell family at Duntroon on the Limestone Plains. There he studied the pastoral practices of a pre-scientific age. Remote as he was from libraries, laboratories and learned men, his studies perforce were solitary; but he had with him a few books and papers, including – we may fairly guess – one or two by the chemist Liebig. His own particular gifts were curiosity, sharp eyesight and the habit of asking well directed questions. In 1873 he published a short book, or, as he called it with characteristic modesty, a paper: *Grass and Sheep Farming: A Paper Speculative and Suggestive*. His purpose, like his prose style, was simple – to put the idea into the heads of practical man that science, if only they would try it, could be of some service to them.

The early 1870s were a period of abnormally high rainfall and high mortality among the flocks. Every practical man could see a connection between wet weather and such diseases as foot rot or liver fluke; but Farrer asked questions about the how and the why of this connection. He could not produce the answers; but he believed that patient scientific research could produce them – or some of them. Chemical research, he suggested, was the immediate need. In successive chapters of his treatise he considered the chemistry of animal nutrition, of plant nutrition, of the nutrient soil. In Germany, eminent men were devoting their lives to the exploration of these fields of study; but in New South Wales nobody, it seemed to Farrer, paid any attention to them.

> If the pastures of this country had fallen into the hands of the Germans, there is little doubt that Pastoral Chemistry would, by this time, have been recognised and cultivated as a distinct branch of science; and as for Agricultural Chemistry, there would be colleges for teaching that in various parts of the colony.

Three decades went by before the parliament of New South Wales voted a few thousand pounds for the purposes defined by Farrer. Six decades went by before Farrer's agenda for research became the agenda of C.S.I.R.O.[1]

So the man on the land still had to discover the hard way the things that he could and could not profitably do. In Monaro, he discovered that it did not pay him to grow grain. That discovery was unexpected and unwelcome. Throughout the 1830s and 1840s, every squatter had maintained his own cultivation paddock and had raised wheat for grinding in his own hand-mill and baking in his own oven. From the mid-nineteenth century onwards, enterprising men like Stewart Ryrie and William Jardine of Jindabyne, Alex Montague of Cooma and John Geldmacher of Nimmitabel found commercial milling profitable. As the railway approached Cooma in the late 1880s, cultivators and millers could hardly wait for the day when flour would be flowing from Monaro to the great market in Sydney. But the opposite happened. The railway brought to Cooma better and cheaper wheat from the Riverina. Year by year the wheat acreage of Monaro shrank until it approached the vanishing point.[2]

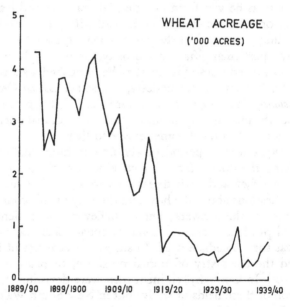

Fig. 1 Wheat acreage, 1889–1939

[1] See pp. 157–61, below. C.S.I.R.O. denotes the Commonwealth Scientific and Industrial Research Organisation.
[2] Perkins, 770, 872, 1038, 1063, 1102, 1650, 1694, 2084, 2365. John Geldmacher built a monumental stone mill with wooden sails. He was not allowed to use the sails on the ground that their shadows would frighten horses. Instead, he used horse power!

It seemed for a short time that dairying had better prospects. During the 1890s there occurred a quite dramatic climb in butter production, not only on small holdings but also on some large properties.[1] Butter factories began to be a fairly common feature of the Monaro landscape. For thirty or forty years, the industry had its ups and downs. But the long-term trend was down.

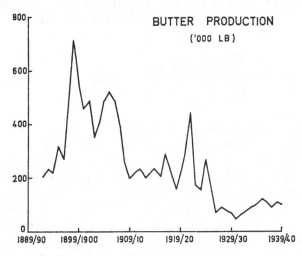

Fig. 2   Butter production, 1889–1939

So once again, environmental hindrances frustrated commercial hopes. Monaro people could no more compete with their neighbours to the east in producing butter than they could compete with their neighbours to the west in growing wheat. If they were to survive and thrive, they would have to do so on their earnings as producers of wool and meat.

Even within this limited range of production their performance during the 1860s and 1870s held a low rating. District Surveyor Betts reported 'the total absence of improvements on the tableland of Monaro'; Crown Lands Commissioner Cooke endorsed that caustic comment.[2] To be sure, the preoccupations of both officials were slanted; Betts was reporting the absence of small agricultural holdings; Cooke was reporting the low rate of investment in water storage. But why should Monaro people build dams or sink wells when their rivers and creeks were still providing as much water as they needed? Cooke saw the point of that question; but

[1] In the late 1890s Bolaro, a property of more than 30,000 acres, passed into the possession of the Scottish and Australian Investment Company. An innovating manager envisaged a kind of Métayer system, under which carefully selected dairy farmers would be established on 7,000 acres of river flats to produce milk for delivery to the estate's butter factory.
[2] V.P. Leg. Ass., 1877–8, vol. 3, p. 384; *Occupation of Growth Lands, Annual Report, 1879*, pp. 150–1.

fencing, he said, was a different matter. Monaro needed fences as much as other districts did, but was exceptionally ill-provided with them.

Superficially, that observation was true; but if the officials had looked beneath the surface they would have seen the start of an immense investment in fencing. Its upward surge is documented in the account books of James Litchfield, the letter books of H. T. Edwards and the diary of Edward Pratt. From 1875 onwards Litchfield regularly recorded large payments for the erection of 7-wire boundary fences. His rates of pay were as follows:

Splitting posts	25*s*. per 100
Carting them into line	20*s*. per 100
Erecting them	4*d*. per post
Boring them	15*s*. per 100
Running the wires	£6 per mile

By the 1880s, Litchfield was hiring contractors to do the complete job at the rate of £60 per mile. Edwards did not stick so consistently to one technique – some of his fences had only three wires – and his payments per mile were generally lower than Litchfield's; but in one way or another he covered as many miles. Pratt started his big burst of fencing in 1877 and maintained it non-stop throughout the following six or seven years. Starting with an order for 12 tons of wire in October 1877 he kept his purchases ahead of rising prices and kept the neighbouring selectors profitably at work on the tasks of timber getting, post splitting, wire straining and the rest.

In the mid-1880s all three men steeply reduced their investment, partly because the years 1885–7 were a period of falling wool prices and low rainfall, partly because they had by now achieved their first objective. That objective was survival. The owners of cattle and sheep could not survive except by converting leasehold into freehold; they could not hold their freehold secure except by fencing its boundaries. No more could they manage it competently if they left it unfenced; Pratt erected his first fence to keep Smith's sheep off his land. From innumerable petty decisions of this kind a more positive objective took shape. To fence the land was to improve it. The work of improvement could not and did not stop with the erection of boundary fences; every improving proprietor wanted fenced paddocks. By 1880 Pratt had a 'rain paddock' on Cooma Creek, a large paddock with seven gates on the Peak, a cultivation paddock, a hay paddock, a woolshed paddock and several small breeding paddocks. On 7 May 1881 H. T. Edwards told his bank manager that the Burnima Estate was 'all newly fenced and divided into small sheep paddocks'. Paddocking by now was well established as the new orthodoxy of pastoral practice. In a much-quoted report of 1875, the Chief Inspector of Stock had enumerated its advantages as follows: the country will carry one-

Improving

third more sheep; the wool will be longer and sounder; the feed will be cleaner; the lambing will be better; the sheep will increase in size; they will have fewer diseases; they will live longer; the expense of working a station will be cut to one quarter of what it had been in the days of shepherding; the owner will be able to devote most of his time to improving his sheep, instead of spending it in attempting to manage a lot of careless shepherds and hut-keepers.[1]

The author of this forceful advocacy, Alexander Bruce, was a shining exception to the general rule that the man on the land had to fend for himself without any administrative or technical aid from the government. Bruce was a farmer's son from Aberdeenshire who had come to Australia in 1852 with no scientific training and no more than a smattering of legal and commercial training. He tried to farm, was bilked by his partner, sold the remnant of his assets and embarked on a career in the public service. From the early 1860s to the early 1900s he was the administrative head of the Livestock and Brands Division of the Department of Mines.[2] There he did for the whole of New South Wales what William Bradley had done for his twenty stations in Monaro; he stamped out the most dangerous diseases of cattle and sheep. By instituting a strict system of quarantine, he greatly reduced the risk of new diseases entering the colony from its colonial neighbours or from overseas. All in all, he was a great administrator. He was also a tireless propagandist for improvement – improvement of the stock routes, of markets at home and overseas, of the quality of the colony's flocks and herds. *How* to improve their quality the pastoralists still had to find out for themselves. Alexander Bruce could give them no advice on breeding policy, because he was untrained scientifically and had to support him not even one scientific research worker.

In the mid-nineteenth century, breeding policy was a highly controversial subject. Imported rams and cross breeding had been the recent fashion; but it produced by reaction a fanatical propaganda for the pure Australian merino. The only thing that mattered, the propagandists asserted, was fine wool of long staple and abundant yolk in a dense fleece; sheep of large size were not really wanted – or, if graziers did want them, the way to get them was by feeding, not breeding.[3]

One wonders whether this spirited propaganda penetrated Monaro. If it did, James Litchfield rejected it. He made up his mind to breed for size, constitution and a medium fibre. In 1865 he established a stud on progeny of Rambouillet ewes. In 1881 he started to import from Tasmania merinos, mostly of Saxon descent, which he selected for their size, constitution

[1] V.P. Leg. Ass., vol. 4, pp. 517 ff.
[2] Originally, the Department of Lands. Transfer of the division to the Department of Mines took place in 1874.
[3] See e.g. James Jordan, *The Management of Sheep Stations* (Melbourne, 1869), and John Ryrie Graham, *A Treatise on the Australian Merino* (Melbourne, 1870).

and medium wool. In the early 1890s he bought from Messrs Millear and Austin of Wanganella a famous ram named Sovereign. At first sight, that purchase might appear a new departure of policy; but the Wanganella stud, like his own, had a strong Rambouillet infusion. His object throughout was to breed large animals of strong constitution with fleeces of medium, long-staple wool.[1]

In establishing a strain of sheep so well adapted to the hard environment of Monaro, Litchfield was dependent for the first ten years on his shepherds. They were not at all the lazy and incompetent workers denounced by Alexander Bruce. Litchfield paid them well, supervised them closely and gave them better equipment than had been customary in the years of squatting; in return, they served him faithfully and efficiently.[2] Even so, his decision in the mid-1870s to make a start with fencing was a big step forward. When the boundary fences were finished he subdivided his land into paddocks ranging from 1,000 to 2,000 acres – a size appropriate at that time to progressive flock management in his district. Until approximately the mid-1880s he continued to breed sheep chiefly for his own flock: in 1884 a flock of 15,000 sheep on 20,000 acres of land produced a wool clip worth £3,180 and earned additional income from meat sold to the local butchers and from rams let on hire during the mating season to neighbouring graziers. For some time past, however, some neighbours had been choosing to buy rams rather than to hire them. By the 1890s, progeny of the Hazeldean stud were in strong demand, no only within Monaro, but far beyond its confines.[3]

The other studs of Monaro were established a good deal later; but from the 1880s onwards many graziers were improving their flocks. In the Bombala district, H. T. Edwards was getting excellent results from his pure Merinos; half way between Cooma and Bombala, William Jardine of Curry Flat was managing a good flock strongly infused with South Australian blood: other successful flockmasters – Evans of Kiah Lake, Harkness of Lincluden, Pryce of Woolway, Locker of Adaminaby – were winning prizes year by year at the Cooma Sheep Show.[4] Good quality fleeces of 7 to 8 lb or even heavier were becoming quite common in Monaro.[5] The progressive diffusion of improved pastoral practices was hap-

[1] *Past. Rev.* xx (April, 1910): an informative article on the Litchfield family and the Hazeldean stud.
[2] The names of these good shepherds are on record. Their wages were £35 a year with rations and a bonus of 20s. per 100 lambs marked in their flocks. They no longer had to shift hurdles about, but had spacious sheep yards walled with the stone which was in good supply on Litchfield's land. Most of the walls had been built by Chinese workers.
[3] In the early years of free selection Litchfield had made his home at Springwell; in 1883 he had a new house built at Hazeldean, a few miles closer to Cooma. The origin of the name is not on record.
[4] Perkins, 1659, 2085.
[5] Cf. H. T. Edwards, 5 December 1886, 'I finished shearing last Friday 30,000 sheep shorn

Improving

pening quite naturally as neighbours observed and imitated the innovators. For example, Pratt was a frequent visitor to Litchfield and an enthusiastic admirer of his rams and ewes, his yards and buildings. Similarly, Jardine visited Pratt and admired his machinery.[1]

Let us spare a paragraph or two for a Pratt's eye view of improvement in the second meaning which Macaulay gave to the word – not the quantities of production, but the quality of life.[2] In the art of flock management, Pratt was a pupil of his neighbour Litchfield; but in the gardener's art he needed no teacher. He had inherited a garden from the Wallaces and started at once to improve it; in his very first week on Myalla he pruned the neglected currant bushes and planted their suckers in a new bed; he planted chestnuts and walnuts at some distance from the house; closer in he planted carnations and pinks. Throughout his life, his favourite medicine in time of trouble was to dig a new flower bed or to make new plantings in an old one; at Christmas and Easter he used to drive into Cooma with great masses of flowers to give to the Church decorators of the various denominations. He was just as much house-proud as he was garden-proud. He liked the house which he had inherited from the Wallaces;[3] but he lost no time in improving it according to his own taste; he combed the second-hand furniture shops of Sydney for marble mantle-pieces, wooden pilasters and similar embellishments. As year followed year, the house took the imprint of the man's style, and the style of his period. And of course it grew bigger as Pratt built new rooms to meet the needs of his growing family.

In the early years, the education of his children – and his neighbours' children too – was a constant worry to him. Only a few weeks after he had bought Myalla he wrote in his diary: 'Sandy McDonald is anxious to get a school in the neighbourhood. I will go in for this heart and soul.'[4] There was a school at Cooma, but the schoolmaster gave up teaching and took to selling grog.[5] One assumes that he was replaced; but the Myalla children could not possibly ride 17 miles to Cooma every day and 17 miles back again. Home teaching by Mrs Pratt met the immediate needs of the master's children; but nearly all the others were growing up illiterate. In 1874 Pratt engaged a schoolmaster at his own expense; but the man was prone to delirium tremens and too often absent without leave in the

turning out 523 Bales Wool averaging 9 lb 1 oz per sheep. I do not think this return has often been beaten in the Colony.' However, the 1886 shearing season in the Bombala district was exceptionally good.
[1] Pratt, 1 February 1883, 3 July 1878, 7 January and 3 June 1880, 31 December 1889.
[2] See p. 72, above.
[3] In 1866 the Cooma firm of Potter and Scarlett had built new houses at Coolringdon, Dangelong and Myalla for the three Wallace brothers.
[4] 18 May 1872. Sandy McDonald was one of the Myalla shepherds.
[5] 22 January 1874.

Cooma lock-up.[1] Pratt pulled all the strings he could to get a government school for Myalla and at long last he got it. It opened in 1881 with a woman teacher and twenty-one pupils. Pratt's handyman, Snowden, had made the school furniture. Snowden and his wife appear frequently in Pratt's diary; he was a willing and versatile worker; she was a good, but an ailing housewife. Mrs Pratt performed the duties of unofficial district nurse to the Snowdens and, no doubt, to other families on and around Myalla.

As we have seen, there was strife on Myalla; but there was also kindness and a neighbourly solidarity. Cricket matches were arranged between Myalla and Hazeldean. Cooma was not too far distant for the occasional jaunt. By the 1880s, Cooma had grown to the dimensions of a well-provided country township, with its own court house, land office, hospital, school, newspaper – not to mention its pubs and shops; at Hain's store the country families could buy tools, kitchen ware, furniture, bedding, clothes, groceries and almost anything else they needed. Moreover, Cooma was as well supplied as any other small colonial town with churches, playing fields, a show ground, a race course – but not, as yet, a library; that was the weak spot. Even so, the Monaro community of country people and town people appeared to be making good progress during the 1870s and early 1880s along a fairly wide front of improvement – and self-improvement.

But the economy of Monaro had a narrow base. It was vulnerable to drought and to falling prices for its staple products, wool and meat. On 27 April 1886, H. T. Edwards wrote a regretful letter of dismissal to one of his workers – 'I am sorry to say the bad times have obliged me to reduce my expenses and do with fewer men.' Later that year the rains came and Burnima produced its record wool clip; but prices continued to fall and the drought returned. On 17 November 1888, Edwards wrote, 'I have 16,000 sheep and a lot of cattle away, but it is *too* late. The grass is gone.' On 29 September 1889 he looked back gloomily on the bad times of 1885–8: '...during those years there was the greatest depression ever known throughout the whole world and Australia did not escape, very few stations paid their expenses.' It was as well for his peace of mind that he could not foresee the far worse time that persisted year after year during the 1890s.

'Bad times in Monaro' would make a rewarding project of historical research. The approach could be from various angles – of the worker who lost his job and failed to find another; of the storekeeper who gave the worker groceries on credit but had also to safeguard his own solvency; of the grazier who could barely cover his working expenses, let alone the interest due on his mortgages. A social historian might choose to focus his inquiry on the unemployed workers and the compassionate storekeep-

[1] 20 November 1874, 5 May and 9 September 1876.

ers; but an economic historian would put his spotlight on the plight of the graziers. Profit or loss on their stations constituted the multiplier of prosperity or hardship among the workers, the storekeepers and all other sorts and conditions of men in Monaro.

As a class, the graziers had to carry the cost of the large sums which they had borrowed to make their stations secure and improve them: in good economic weather they carried that cost easily enough, but they found it back breaking when the weather turned dirty. Even so, most of them pulled through. A story used to be told of the banks and pastoral companies taking over the ownership and management of nearly all the stations of New South Wales; but that story has now taken its proper place in the long list of Australian legends.[1] In the sorely afflicted Western Division the story came fairly close to the truth; but in the Eastern Division it was far from the truth. In Monaro, foreclosures did sometimes happen and, if time permitted, it might be worth while to count and map them. For the present, however, the twenty Bradley properties must serve as a sample. From the 1860s right up to the First World War, all except one remained continuously in private ownership.[2] If the sample is at all representative, what we have to explain is not a high casualty rate but a high survival rate. Here are some explanations which come readily to mind. First, 'the twenty years' turnover' of ownership which people took for granted on the western plains was not the custom of Monaro; in the slump of the 1840s the land speculators had fled, but many of the land users stayed put and continued to stay put through good times and bad until they achieved at least a rough and ready understanding of their environment. Secondly, that environment was moderately steady; it offered no glittering prizes and threatened no irremediable disasters; it favoured the steady man. Thirdly, Monaro escaped the terrible drought which nearly everywhere else followed the slump of the 1890s.[3] Fourthly,

[1] See N. G. Butlin's magisterial article, 'Company Ownership of NSW Pastoral Stations, 1865–1900', in *Hist. Stud.* IV (1950), no. 14.
[2] For the single and absurd exception of Dangelong, see p. 92, above.
[3] *Annual Rainfall at Bukalong (average 1958–69 = 25.45 in.)*

Year	Inches
1893	35.53
1894	22.14
1895	15.42
1896	27.55
1897	27.31
1898	20.37
1899	25.36
1900	27.03
1901	19.69
1902	24.65
1903	17.46
1904	17.18
1905	22.36

Possessing the land

the rabbits were latecomers to Monaro and, on the basaltic plains at least, were dangerous but not invincible invaders. Whereas properties in the Western Division suffered irreparable damage, the pastoralists of Monaro could have said, as William Bradley had said half a century before, that they had suffered no diminution of their real resources.[1]

They could have said even more than that, because they could now see the promise – hesitant though it still was – of a new deal in terms of scientific aid. Hitherto, the government of New South Wales had remained just as much a laggard in science as it had once been in surveying;[2] but in 1891 it established a Department of Agriculture. That title, no doubt, was too pretentious; agriculture remained until 1908 a branch of the Department of Mines – an inappropriate, and too often a niggardly stepmother to poor Cinderella. In 1891, the argicultural budget was cut to £10,000. On this pitiable income a few devoted men performed miracles; they published an *Agricultural Gazette* and wrote most of its articles; they founded Hawkesbury Agricultural College, with other colleges soon to follow; they established experimental and demonstration stations in various districts of New South Wales. The intention of these ventures was deliberately practical; but some members of the small staff somehow or other found time for original scientific research. The Principal Scientific Officer, Dr Nathan Cobb, was a German-trained biologist of distinguished achievement over a wide field. In 1898 he persuaded William Farrer to join the department at a salary of £350 per annum. For ten years past, Farrer had been conducting experiments in the cross-breeding of wheat. He had conducted them in his own laboratory on his own farm at his own expense. Their outcome was an immense augmentation of opportunities and incomes on the land. When Farrer died in 1906 his admirers launched a Farrer Memorial Fund; but subscriptions began to taper off before £200 had been collected. The people of New South Wales were not mean; they were unimaginative. It took them a long time even to begin to understand how it was possible for them to be so deeply in debt to such a quiet hero.[3]

In the early 1870s Farrer had studied grass. In the early 1880s he married into the de Salis family, graziers at Cuppacumbalong in the Murrumbidgee valley. His own farm, Lambrigg, was predominantly a grazing

A succession of dry years did not occur on Bukalong until *after* the climax year (1902) of the disastrous drought in most of N.S.W. However, since Monaro contains a good many microclimates (see map 5, p. 13) the figures cited above for Bukalong must not be taken as proof that the Bredbo or Dalgety districts – for example – enjoyed comparable immunity from drought.
[1] See p. 78, above.
[2] See p. 90, above.
[3] In writing this paragraph I have been in debt to an unpublished paper, *The Early Years of the Department of Agriculture in New South Wales*, by Mr C. J. King, until recently the department's Chief of the Division of Marketing and Agricultural Economics. Today, there are Farrer Scholarships and a Farrer Memorial Lecture.

Improving

property. His friends and neighbours were graziers. For all these reasons one feels surprise that he made wheat, not grass, the object of his research. The graziers of New South Wales stood almost as much in need of science as the cultivators did. Even the best of them still had almost everything to learn about the sciences and art of pasture improvement.

Up to the end of the nineteenth century, their struggle for more and better pasture had produced one innovation – the fenced paddock – which everybody applauded; but another innovation, the ringbarked tree, became the occasion of controversy. So long as people believed that trees brought rain, ringbarking had been taboo; but from the 1850s onward the opinion gained ground among men of science that trees produced no modification of the macro-climate. In 1873, Farrer forcefully advocated ringbarking as the cheapest and best method of making available for the growth of grass the soil, water and sunlight hitherto pre-empted by the trees.[1] At Ginninderra, only a few miles distant from the place where Farrer wrote his treatise, a selector named Samuel Schumack attacked with his American axe almost every tree on his land. Neither then or later did he feel the least doubt that trees were his natural enemies – although he might well have done so, when gully erosion began to spoil his land.[2] Farrer, to do him justice, had never advocated indiscriminate ringbarking. Gradually the opinion gained ground that very careful discrimination was called for before the axe was laid to the bark of the tree. Good pasture was not the invariable successor to a stand of eucalyptus; the successor might be poor grass, or impenetrable scrub, or a landslide.[3] Thoughtful people began to see the necessity of establishing an alternative soil cover as good as, or better than, the trees that had been destroyed. Thus, by a natural transition, men of science and practical men began to envisage measures to improve the Australian pastures.[4] A good deal of thought was given to the possibilities of harvesting and sowing the native species; but that path led to a dead end. If ever the job were to be done, exotic species had to do it.[5]

In the last phases of pasture degeneration and in the opening phases of pasture improvement, the experience of Monaro did not deviate far from the norm prevalent throughout south-eastern Australia. On the basaltic downland, ringbarking was seldom practised, because there were so few trees; but in the timbered country to the west and to the east it was commonly practised, with the same mixed consequences as elsewhere. In

---

[1] King, *Early Years of the Department of Agriculture*, pp. 36, 65.
[2] *Autobiography* (A.N.U. Press, 1967), pp. 28, 109–11. After heavy rain in February 1883 Schumack cleaned out 5 ft of silt from each of two dams which he had made recently.
[3] The pros and cons of ringbarking were much discussed in *Past. Rev.* (e.g. 15 August 1895, 16 August 1896).
[4] *Past. Rev.*, 15 August and 15 September 1891, 15 May 1894, 15 November 1897, 15 April 1898, 15 April 1899.
[5] For the reason, see p. 151, below.

every kind of country, the graziers were finding it hard to cope with the proliferation of noxious plants, both native and imported. Increasingly, they sought the remedy of summer grazing in the high country. The alternative remedy – pasture improvement on the tableland – was foreshadowed in the 1860s, when lucerne was first established on William Bradley's land.[1] To some people, that seemed a rash innovation. Archdeacon Druitt, as trustee for Boloco during the minority of its heirs, planted quite a lot of lucerne; but when the heirs came of age they sued the Archdeacon for administering his trust extravagantly.[2] Nevertheless the experiments with exotic plants continued. By the 1880s, James Litchfield had some hundreds of acres under lucerne on the alluvial flats of Cooma Back Creek and the Brothers Creek. Edward Pratt made a rye grass paddock; other graziers made small sowings of lucerne, rye grass, cocksfoot or prairie grass.

Thus did the idea of pasture improvement find hesitant acceptance here and there in Monaro. Prospects of an advance on a wide front did not come into sight until 1925, when the first sowings of subterranean clover with superphosphate were made on Delegate Station. Even then, as will appear later, the hopes of a big leap forward had to be deferred.

[1] See p. 80, above.
[2] Information from Mr H. A. Rose.

4. White men in the high country: the first hundred years

Since this book began to be written a firm step has been taken towards the archaeological discovery of Monaro. As one investigation leads to the next, scientific knowledge of the Aboriginal past will progressively supersede the guesses that were hazarded in an earlier chapter. Nevertheless, some of the statements made there went beyond guesswork. We do not yet know how many thousands of years the Aborigines had been in the high country before the white men invaded it; but we do know that they had been there for many thousands of years. We also know that they were summer visitors, coming and going with the Bogong moths.[1]

Beyond that, we can feel nearly certain that they did not seriously damage the high country by fire. For reasons that were explained earlier, they were inveterate incendiaries on the tableland; but in the high country they had no motive for starting big fires. Bogong moths, packed tight in rocky shelters, were their quarry. To prepare the moths as food they needed only small fires, carefully reduced to ash. In most years – although not in all – the moths and the men would be starting to leave the high mountains before the fire danger there became extreme. This is not to say that high-altitude fires never got away or that low-altitude fires never worked their way upwards; scars in the sap rings of very old trees are proof to the contrary. These scars, nevertheless, are of infrequent occurrence, in comparison with those that began to scar the trees soon after the white man arrived.[2] We may accept as broadly applicable to the entire massif what a uniquely well informed forester wrote forty odd years ago about the alpine ash country at the 3,000–4,500 ft contour levels.

Before the advent of white men in Australia, fires did undoubtedly occur, they

---

[1] See pt. 1, 2, above. Josephine M. Flood is now undertaking an archaeological survey of Monaro's tribal territories. Sites which she has already located are yielding and will yield rich material.

[2] The tree ring evidence has been examined by A. B. Costin of C.S.I.R.O. and by two scientists formerly employed by the S.M.A., J. E. Raeder-Roitsch and Marie E. Phillips.

were lit by the blacks on the grass country of the plains and no doubt some of them reached the mountains; but the interval between successive severe fires was great and this fire-sensitive species [alpine ash] was able to develop.

Since white settlement began, fires have increased in number until at the present time severe fires occur every five to ten years.[1]

We may reasonably conclude that the first white stockmen saw little man-made damage when they drove their cattle into the high country.

Nor did they encounter any geographical barriers worth mentioning, as they made their way from the Goulburn plains to Gundagai, up the Tumut valley and over the 4,000–5,000 ft contours to the headwaters of the Murrumbidgee and Eucumbene. There they found good summer pasture in a subalpine landscape of open forest and grass hollow; from there they could move still further south along the flattened range which led through snow gums and snow grass to alpine herbfields above the tree line. What they did not then understand was the extreme danger of the violent blizzards they might encounter if they moved into the high country too early in the summer, or stayed there too late.

The grasslands around Kiandra used to be called Gibson's Plains, after Dr Gibson of Goulburn. His stockmen were among the first to arrive and among the first to suffer disaster. In the Sydney *Monitor* of 25 October 1834 a young squatter reported a terrible blizzard which had overwhelmed one of Gibson's stockmen and destroyed many cattle.

> I may consider myself a fortunate fellow [the young man wrote] as I removed four hundred head of cattle from a station near to Dr Gibson's and Mr Palmer's only one day before the storm fell; had I not got them off as I did, the whole must have inevitably perished. I endangered the lives of my brother, myself, and servant, in consequence of travelling so far over those mountains in search of a station for our stock . . . we could observe a very deep ravine falling to the westward, as far as the eye could carry, at the bottom of which we observed a river: might not this be the source of the Hume? To the right and left of this ravine were mountains which appeared perfectly barren; to the east were smaller ravines falling into the Snowy River.

The young man must have made his way many miles south from Gibson's Plains. Imprinted on his memory was the superb view which Strzelecki saw six years later from the steep escarpment of the westward facing mountains.

Danger was no deterrent to the squatters who sought relief in the high country for the animals that were starving on their over-stocked and drought-afflicted stations. Among those squatters was Terence Aubrey Murray of Yarralumla. In the New Year of 1839 he made his way with a

[1] B. U. Byles, *Report on the Murray River Catchment in New South Wales* (Bulletin No. 13 of the Commonwealth Forestry Bureau, 1932), p. 22. Cf. p. 30, 'I am quite safe in saying that 99 per cent of the fires in the mountains are lit by human agency.'

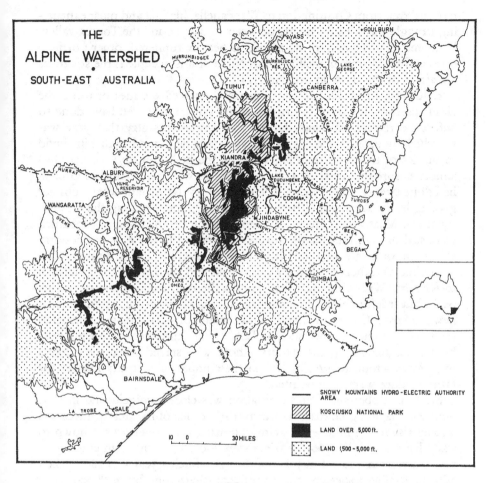

Map 13

party of six over the Brindabella mountains, down into the valley of the Goodradigbee and up again to the headwaters of the snow-fed rivers. There he established outstations for his cattle.[1] By this time, the northern stock routes to the high country were coming into regular use; but little use was being made as yet of the routes leading westwards from the tableland of Monaro. In 1840, when Stewart Ryrie went exploring with his compass, sextant and sketch book, he praised the good pasture under the snow gums and above the tree line; but he reported no cattle or sheep grazing on it.[2] Fifteen years later, when W. A. Brodribb set out on his four

[1] Gwendoline Wilson, *Murray of Yarralumla* (Melbourne, 1968), pp. 104-10.
[2] On Stewart Ryrie's 'amateur survey' of the Monaro Squatting District see p. 39, above.

months' trek from Coolringdon to Wanganella, his first and most exhausting struggle was with the ranges between Jindabyne and the Tooma valley. '...some of this country,' he noted, 'is heavily timbered, other portions open and rocky, but all well grassed. None of it is occupied; the squatters are afraid of the snow in the winter.'[1]

Evidently, Brodribb had not hitherto conceived the idea of using the alpine pastures for relief in dry summers; but he may well have done so before he reached Wanganella, for the western country that year was scorched by drought. We may feel sure that he carried with him vivid memories of the good feed and water of the high country; we may feel almost as sure that he shared those memories with his friends, as Gibson had done twenty years before. The day was not far distant when summer grazing in the Alps would become common practice among the western squatters. Meanwhile, pressures were building up in Monaro. Stockmen were still driving their cattle to summer grazing in the Kiandra country and some of them were on the lookout for gold. In November 1859 the Pollock brothers made their lucky strike and sparked off the gold rush to Kiandra. Within the next few months the rush reached and passed its peak; but it left behind some permanent consequences, which included a new road from Cooma to Kiandra and a scatter of starveling settlements in sheltered pockets of land, such as Lob's Hole, and even on open slopes higher up. Inevitably, the settlers – those who stayed when the diggers went away – took to grazing, when their hopes of a steady market for farm produce were disappointed.[2]

Sir John Robertson's land legislation was closely coincidental in time with the Kiandra gold rush. As we saw earlier, one of its consequences was a steep rise in the number of animals grazing on properties cut down to size.[3] By the middle 1860s, graziers were looking to the high country to save them from disaster in years of drought. Such a year was 1865, when William Bradley's sheep manager for the northern runs drove 48,000 sheep and 2,000 cattle into the mountains. 'My idea,' he recalled in his old age, 'was quite untried at the time although now it is one of the commonplace incidents of every station.'[4] The first statement of that sentence may be open to question, but the second is well supported by frequent press reports in the summer seasons of the 1860s, '70s and '80s. For example, in 1869 and 1871, graziers both small and large – including James Litchfield – had thousands of sheep in the high country and suffered heavy losses when winter weather set in early; in 1885 the weather was 'fearfully dry

W. B. Brodribb, *Recollections of an Australian Squatter*, p. 82.
[2] Good material on the gold rush and its sequel will be found in D. G. Moye, *Historic Kiandra* (Cooma–Monaro Historical Society, 1959). Ring-barked trees remain as evidence of the attempts at land settlement made during and after the gold rush.
[3] See p. 180, above.
[4] W. Davis Wright, *Canberra* (Sydney, 1923), p. 92.

and hot' and the Snowy Mountains were 'overcrowded with sheep'.[1]

The earliest recorded criticism of this relapse to pastoral nomadism appeared on 9 November 1869 in the *Sydney Morning Herald*. It will repay quotation at some length.

> There are now about half-a-million of sheep and lambs in Monaro (in 1868 there were 537,674); and I am informed, on good authority, that 100,000 of these are in the excess of the number it can usefully and properly carry. The owners seek to compensate themselves for 'the fall in wool' by multiplying, to an undue extent, the backs that carry it . . . During a favourable season like the present, when the district is covered, from one end to the other, by rich waving grass, there is feed enough for the entire half million; but . . . when there is a drought – as there was last year and may be in any year – the district must prove unable to support such numerous flocks; they will die in thousands or must be driven 'to the mountains', in their weakened state, with much labour, and at great risk and certain cost, as was the case last year. I should have supposed the wiser policy would be to improve the breed, and quality of the wool, to the highest attainable point of excellence, to fence the runs so far as tenure, means and other circumstances will permit; and where possible to do something in the way of introducing English grasses; for one-half of the extent of land, and one-half of the number of sheep that are now required, or thought to be required, for a station, would then yield as much as all now do; and the cost of management would be reduced by more than one half.

The tone of that advice was rather too superior. Innovating proprietors, as we have seen, were already committing themselves to a heavy invest-ment in fencing and were doing their utmost to improve the quality of their flocks. The day of pasture improvement, however, was still far dis-tant. Until it dawned, even the most progressive managers remained heavily dependent upon the mountain pastures. By the late 1880s, H. T. Edwards had lifted Bibbenluke to a carrying capacity of one sheep to the acre; but he would have been unable to do so had he not lightened the pressure on his paddocks by sending 10,000 to 20,000 sheep every summer into the Kiandra country.[2] The Litchfields were making similar use of the alpine pastures dominated by the Brassies and Kerries ranges, due east of Cooma. How they did so will repay a brief inquiry.

In 1891 James Litchfield retired, after making an equal division of his land between his four sons. They decided to pool their resources and manage them as a common concern.[3] In the main they continued their

---

[1] Perkins, 1016–17, 1681, 1729, 1784, 1813, 1188–9. In excellent records kept by the Rutledge family of Carwoola and Gidleigh there are detailed accounts of the drafting of many thousands of sheep during the 1880s to Happy Jacks near Adaminaby. (Information from Martha Camp-bell, *née* Rutledge.)

[2] Letter Book (20 November 1888) of H. T. Edwards in Bibbenluke and Burnima Papers, N.L.A.

[3] The partnership of Litchfield Bros comprised Arthur, the eldest son, on Hazeldean; Edwin on Woodstock; Owen on Springwell; Frederick on Matong, south of the Snowy. In 1903 the last-named sold Matong and moved to Queensland.

father's policies; but they made one new departure to meet the new opportunities arising from the improvement of communications – in particular, from the extension of the railway to Cooma. This gave them access to the Sydney meat market and made it worth their while not only to maintain the stud and the breeding flock but also to buy young wethers of good quality. They bought them off shears in the early spring and had them brought in by road to graze with the aged sheep which were to be fattened during the summer months and sold in the autumn. To make room for them, the older wethers (from two to four years old) were moved into the high country. This strategy served two purposes. First, it reduced grazing pressure on the home paddocks and thereby improved the prospects of selling prime stock at good prices. Secondly, it improved the prospects of a profitable wool clip by shielding many of the best wool growers from the menace of corkscrew grass. Summer grazing in the mountains thus became for the Litchfields an integral part of flock management.[1] In years of extreme drought they might send cattle as well as sheep; if so, they would follow the Monaro custom of confining the cattle to the steeper, rougher country.[2]

Up to the last decade of the nineteenth century there was not even the pretence of official control over grazing in the high country.[3] Control began when the Lands Department, hungry as ever for revenue, made provision for snow leases and some related tenancies in the Crown Lands (Amendment) Act of 1889. Four years later, Richard Helms reported that 81,000 acres of summer pasture adjacent to Mt Kosciusko had been divided into twenty-two snow leases. Until then, Helms said, the mountains had been 'free country' for anybody's cattle and sheep – and, of course, for the brumbies.[4]

With governmental regulation there began a struggle for the possession of scarce resources. We shall have to follow the course of this struggle closely; but before we start doing so we may allow ourselves the luxury of a digression. The men and the horses of 'the free country' are legendary.

[1] Information from J. F. Litchfield. By the beginning of the twentieth century the pest of corkscrew grass (*Stipa setacea* R.Br.) was widespread on the tableland; but in the autumn, when the sheep in the mountains were brought home, the damaging seed had already fallen.

[2] Whereas the alpine pastures in Victoria remained predominantly cattle country, the trend in New South Wales was steadily towards sheep. Cattle were grazed chiefly on the rougher slopes on the south and west, where dingos were still numerous. Cf. Byles, *Murray River Catchment*, p. 25.

[3] To this statement, one exception must be made: the boundaries of a few runs adjacent to the mountains ran ostensibly to the summits: see e.g. the boundaries of Run 49, 'Rock Forest', as gazetted in 1848.

[4] *Agric. Gaz.* (NSW) IV, 450, 'Report on the Grazing Leases of the Mount Kosciusko Plateau'. Brumbies were horses that had escaped into the bush and were running wild.

In Farquhar McKenzie's diary, far back in the first decade of squatting, we meet the prototype of The Man from Snowy River.[1] A folk hero of the camp fires and bush huts, he soon makes his way into print – into narratives of travel, novels, short stories, ballads. In Banjo Paterson's famous ballad he becomes the symbol of Australian manhood.[2]

> He hails from Snowy River, up by Kosciusko's side,
>   Where the hills are twice as steep and twice as rough;
> Where a horse's hoofs strike firelight from the flint stones every stride,
>   The man that holds his own is good enough.
>
> And the Snowy River riders on the mountains make their home,
>   Where the river runs those giant hills between;
> I have seen full many horsemen since I first commenced to roam,
>   But nowhere yet such horsemen have I seen.

Paterson's hero is a shy stripling on a weedy looking animal like an undersized racehorse. He rides where the big men on their big horses dare not follow him; up and down and over the mountains he pursues the brumbies until he corners them, turns their heads for home and brings back with them an escaped colt worth £1,000 which has been running wild in their company. No wonder that:

> The Man from Snowy River is a household word today
> And the stockmen tell the story of his ride.

In Paterson's galloping verse the Wild Colonial Boy comes of age as a self-confident Australian. Paterson was a squatter's son who loved life out-of-doors, moved easily in the best society and lived to a ripe old age. His younger contemporary, Barcroft Boake, was just as much in love as Paterson was with the Australian out-back; but he hanged himself on the bough of a tree with his own stockwhip when he was only twenty-six years old.[3] His ballads are as sombre in theme and tone as Paterson's are sanguine.

> I must onward and cross the River:
> So long mate! for I cannot stay;
> I must onward and cross the River –
> Over the River there lies my way!

[1] See p. 35, above.
[2] A. B. Paterson (1864-1941) was nicknamed Banjo after an old racehorse he was fond of. *The Man from Snowy River* appeared in 1891 in *The Bulletin* and was republished in 1895 in *The Man from Snowy River and Other Ballads,* a slim volume which sold 10,000 copies within a few months. It was not until many years later that Paterson published *Waltzing Matilda,* a ballad equally famous.
[3] Barcroft Boake (1866–92) was the son of a photographer. He earned his living successively as surveyor's assistant, boundary rider and drover. Most of his verse was written in 1891-2 and published in *The Bulletin*. A collected edition, *Where the Dead Men Lie and Other Poems,* was published in 1897 with notes and a memoir by A. G. Stephen.

. . . .
Where is he making for? Down the River
Down the river of slimy bed!
Where is he making for? Down the River,
Down the River that bears him – dead.

Most of Boake's heroes, like Paterson's, are men on horseback; but one
of them, Carl the Dane, is a man on skis – or snow-shoes, as the Kiandra
people called them.

His long, lithe snow-shoes sped along
In easy rhythm to his song;
Now slowly circling round the hill,
Now speeding downward with a will
The crystals crash and blaze and flash;
As o'er the frozen crust they dash.

Davy Eccleston challenges Carl the Dane to a snow-shoe race. The girl
they both love declares that she will marry the winner. The devil joins
Carl on a practice run and gives him supernatural snow-shoes in exchange
for the cross, a gift from his mother, which hangs from a chain around
his neck. Carl wins the race and disappears for ever in the waste of snow.

But now the lonely diggers say
That sometimes at the close of day,
They see a misty wraith flash by
With the faint echo of a cry . . .

The hearty Paterson and the melancholy Boake would have felt the same
surprise had they been able to foresee the day when the high country
would be empty of stockmen in the summer but crowded with skiers in
the winter. Commercialised winter sports are such big business nowadays
that it will be worth our while to explore their humble origins in Kiandra.
Our starting point is a news item of 29 July 1861 in the *Monaro Mercury*.[1]

Kiandra is a rather dreary place in the winter, but yet the people are not
without their amusements. The heaven-pointing snow-clad mountains afford
them some pleasure. Scores of young people are frequently engaged climbing
the lofty summits with snow-shoes and then sliding down with a volancy that
would do credit to some of our railway trains.

Those 'snow-shoes' need to be explained. Some diggers used the Canadian
article – a specimen is still preserved in the Cooma Museum – but the
'demon snow-shoes' of Barcroft Boake's ballad were genuine skis or, as
they are still called in Norway, shis. An alternative name for them in
Kiandra was 'Lapland skates'. We have every reason for accepting the

[1] In the following paragraphs I am in debt to Mr Stewart Jamieson for information and for
references to contemporary newspaper reports and to articles in the *Australian Ski Year Book*.
D. G. Moye has made good use of the same references in *Historic Kiandra*, pp. 62–5.

tradition of their Scandinavian provenance. From strictly contemporary evidence, we know how they were made and what they looked like.

> The skates are constructed of two palings turned up at the front and about four feet long, with straps to put the feet in, and the traveller carries a long stick to balance himself and to assist him up the hill. Down hill they can go as fast as a steamer and up hill, with the aid of the pole, they can make good headway.[1]

In the second winter of the gold rush palings were easily come by, because most of the fences and huts were already tumbling down. Later on, Kiandra people had to make their skis from sawn timber. They made them with multiple grooving on the butter-pat model. Before very long – the precise year is not known – they established the Kiandra Snow-Shoe Club. The name need not bother us: it was a ski club, and one of the first to be established anywhere in the world – although the Swiss, when they went into the snow sport business a generation later, would have smiled to see Kiandra's langlaufers setting out in their Sunday best suits, with starched collars, dickeys and bowler hats.[2]

Charles Kerry, a first-rate photographer who had grown up in Monaro, introduced his Sydney friends to ski-ing and became the moving spirit of the New South Wales Alpine Club. Its members were an influential pressure group. For a time, some of them had hopes of getting the railway extended from Cooma to the summit of Kosciusko; but in the age of motor transport, that would have been superfluous: in 1909 they got a road, with an hotel thrown in. That was the first splutter of the tourist explosion in the high country. In 1920, enthusiasts in New South Wales and Victoria joined forces in the Ski Club of Australia. In 1928, they published the first volume of the *Australian Ski Year Book*. Its editor looked forward to the day when hotels and villages would 'nestle at the foot of every snowy peak' along the 75 miles from Kiandra to Kosciusko.[3]

Meanwhile, the Lands Department had been trying to tidy up the high country. Our immediate task is to identify the main landmarks of its policy – insofar as it possessed a policy – from the early 1890s to the late 1920s. Thereafter, we shall trace the emergence of the idea that land use in

[1] *S.M.H.*, 12 August 1861. Other contemporary accounts speak of two ski sticks instead of a single pole or stick.
[2] The Christiana (Oslo) Club was founded in the early 1860s, the Kiandra club not many years later. In 1927, three or four of the best Kiandra skiers, in the decorous garb described above, competed against the Kosciusko champions (information from Stewart Jamieson).
[3] *Australian Ski Year Book*, vol. 1, p. 3. Cf. pp. 128 ff., an account by Dr Herbert H. Schlink of the first ski tour from Kiandra to Kosciusko. This feat might not have been possible had not the brothers Litchfield agreed to convert a small and primitive hut on the snow lease under Mt Gungartan into a commodious weather-proof 'tin hut'. The only thing amiss with it was – and is – a smoky chimney.

the high country, so far from being the exclusive concern of officials and graziers, is a matter of national concern. This agenda is large and important enough to require a substantial monograph; but here we can spare no more than a few pages to spotlight salient problems as they first emerge during the late nineteenth and early twentieth centuries.[1]

The Act of 1889 had instituted a tenancy called the snow lease; but when the first snow leases were advertised a year or two later the demand for them was disappointing. Graziers who had hitherto paid nothing for their share of the alpine pastures shied away from the newly imposed costs of survey, rent and improvements – the more so as the snow leases ran only for seven years, with the possibility of an extension to ten years. Officials in Sydney believed the resistance of the graziers to be calculated: if tenders for the snow leases were few and far between, the mountains would remain 'free country' for almost everybody. Be this as it may, the Lands Department was failing to collect the revenue it had counted on. To make the shortfall good, it offered alternative tenancies: at one end of the scale, the scrub lease, a fourteen-year tenancy renewable to twenty-one years: at the other end of the scale, the occupation licence or annual lease: in between, the permissive occupancy, a tenure granted by royal prerogative; it ran initially for one year, but was renewable, subject to the Minister's consent, for a three-year period or even a succession of such periods. The bits and pieces of this patchwork were supposedly in accord with environmental variations; but they were more demonstrably in accord with monetary calculations. Administratively, they were a nuisance. Efforts towards a uniform and more manageable system were made between 1909 and 1917, and again during the middle and late 1920s. A main objective of these efforts was to make the snow lease so attractive that the other tenancies could be dispensed with. By an amending Act of 1917, the term of a snow lease was extended to fourteen years, which seemed long enough to offer the lessee a reasonable return on his investment. That calculation proved realistic; during the 1920s snow leases came into strong demand. For the first time, it became possible to curtail the alternative tenancies, excepting the permissive occupancies; these were retained for a time as a flexible instrument of drought relief and as a halfway house where politicians and officials could take shelter, while they were making up their minds what settlement they would make in long term of this or that mountain area. Even so, the permissive occupancies also were on their

---

[1] Here I must offer an explanation. I have in my files a good deal of information on the history of the snow leases. It includes a precise tabulation of particulars of many individual tenancies, made for me by Miss Kaira Suthern from the Lands Department records. It also includes some excellent maps drawn for me by Mr Dan Coward. My original intention was to embody this material in a sizeable chapter; but when I realised that a monograph would be required, I decided to save the material for the use of its author – if and when he appears. The few exploratory pages which follow will not, I hope, queer his pitch.

way out. By the late 1920s the snow lease was well on the way to becoming, at least in everyday speech, the sole term of description for grazing rights in the mountains.

Our task now is to identify in a rough-and-ready way the types of people who were using the mountain pastures from the 1890s onwards under one kind of tenancy or another. In the early decades, big men from the Western Division of New South Wales were conspicuous on the lists of lessees. Throughout the long drawn-out departmental inquiry which started in 1909 tenancies such as the following were frequently recorded.

James Riall   42,821 acres; annual lease

J. and A. M. Pierce
$\left\{ \begin{array}{l} 3,000 \text{ acres; improvement lease} \\ 56,893 \text{ acres; occupation licence} \\ 39,438 \text{ acres; annual lease} \end{array} \right.$

A. B. Triggs
$\left\{ \begin{array}{l} 63,470 \text{ acres; scrub lease} \\ 5,550 \text{ acres; improvement lease} \\ 5,400 \text{ acres; snow lease} \\ 8,807 \text{ acres; occupation licence} \end{array} \right.$

In the main, these big men were buying insurance against drought. We may feel sure that the mountains were crowded with their animals during the great drought at the turn of the century and that the soil cover suffered severe damage during those years. In good years, on the other hand, the westerners sent far fewer animals, if any at all, to the mountains. From the Lands Department records it appears that some of them tried to recover the cost of their tenancies by selling agistment to drought-afflicted graziers of other regions. It appears also that pasture thieves were roaming the mountains: in 1909 a man from Oberne, the holder of a permissive occupancy near Dicky Cooper's Bogong, gave evidence as follows: 'I did not use the land. It was jumped on and the grass was eaten off before I could get my sheep on it.' Just as much a pest were the speculators who owned no land but made high bids for short-term mountain leases so that they could fatten sheep or cattle for the city markets. We can feel sure that these men crowded and damaged the pastures.

Although it would be quite wrong to suggest that western graziers as a class misused the high country whereas eastern graziers used it well, we need feel no surprise that resentment grew among the smaller man of the east against the bigger men of the west. Within the Lands Department, the opinion gradually prevailed – although not without some debate – that a larger share of the leaseholds should be granted to graziers whose home stations were close to the mountains. In a forceful memorandum of 24 November 1924, A. H. Chesterman, the Surveyor-General, made the following points:

1. Scrub leases were bad, because they ran for too long a term and permitted the occupation of too large an area.
2. Snow leases were good, because they ran for a reasonable term (the original seven-year term could now be extended to fourteen years) and stipulated a reasonable area (in 1924 the average area of a snow lease was 6,633 acres).
3. In the allocation of leases, more consideration should be given henceforward to the claims of graziers in the eastern Riverina and in Monaro.

In 1929/30, Monaro graziers were holding 36 per cent of the snow leases.[1] Since the western graziers – whether far distant from the mountains or close to them – were established for the most part on the western slopes, Monaro's share was not stingy.[2]

It remains for us to inquire what use the Monaro men had been making of their mountain grazings. Unfortunately, we are faced in this inquiry with a dearth of written record: the chair bound officials of the Lands Department knew next to nothing about land use in the mountains or anywhere else: in the station records which still survive, references to the high country are sparse. Nevertheless, we can trust the folk memories which tell us – for example – that the men of the Mowamba valley still used the Mount Pilot range, inside the law or outside it, in the same free and easy way in which their fathers had used it. Beyond that, we know with some precision how the Litchfields used the country which rises from the Snowy valley to the headwaters of the Whites and Finn Rivers beneath Gungartan, the Kerries and the Brassies: we even know the names of the shepherds who were working for them until 1910 or thereabouts, when brush and wire fencing made shepherding superfluous. In 1910 the Litchfield brothers dissolved their partnership; but they continued to manage the alpine pastures as a combined operation. Since each brother had the right to apply both for a snow lease and a permissive occupancy, they secured in combination an extensive tract of territory. Around 1910, their combined flocks in the high country totalled approximately 10,000 sheep; but, possessing such ample pasture, they were able to graze it lightly. It was their custom to move their sheep up and down the mountain in prudent accord with the rhythm of the summer season. By contrast, an individual grazier with far less pasture at his disposal was under strong temptation to flog it the whole time that it was free from snow.

In the records of the Lands Department up to about the 1930s, concern for the preservation of the water catchments is conspicuous by its absence.

[1] No comparable figures for permissive occupancies are on record. In 1929 it was ruled that no grazier could hold more than one snow lease; but he was allowed to hold a permissive occupancy in addition to it.
[2] By 1943, Monaro's share had risen to 43 per cent.

White men in the high country

Clauses in the early leases, stipulating the preservation of timber, had primarily a commerical intention. Particular prohibitions against the destruction of vegetation along the banks of streams had some conservationist significance; but not until the late 1920s did leasehold agreements contain any prohibitions against over-stocking or burning. In practice, such prohibitions were ineffective because the Lands Department made no serious attempt to police them. Not until the 1940s did it appoint two rangers – a ridiculously inadequate patrol for so large a territory.

Men of science did not take the lead – as Australians of the mid-twentieth century would expect them to have done – in the movement for catchment protection.[1] Scientific exploration of the high country, after its flying start in the 1850s with W. B. Clarke and Ferninand von Mueller, came almost to a stop during the next three decades. A second start was made in the late 1880s.

Men of Science in the High Country

1885: R. von Lendenfeld: a geographical, geological and meteorological reconnaissance.

1889 and 1893: R. Helms: brilliant probing along a wide scientific and historical front.

1898 and 1899: J. H. Maiden: botanical collection and classification in the von Mueller tradition.

From 1898 continuously to 1902: Clement Wragge: weather observations at Mt Kosciusko (and also at Merimbula on the coast).

1901 and 1907: Edgeworth David: fundamental geological research – among other things, into the evidence of glaciation.

The investigations thus briefly chronicled produced large increments of scientific knowledge. Moreover, they were immensely enjoyable. Von Lendenfeld, for example, enjoyed roughing it in company with the young-er James Spencer, son of 'the king of the mountains', whose exploits are still legendary in Monaro's folklore.[2] Every scientific explorer of that time, with one exception, flung himself with gusto not only into his tasks of research but also into the everyday life of The Man from Snowy River. The exception was Richard Helms. As we saw earlier, he tried to find out how the Aborigines had used the land on which they lived.[3] He wanted

[1] See pp. 166, 170, below.
[2] *Report by Dr R. von Lendenfeld* to the Minister of Mines, 21 January 1885, p. 13. I have had pointed out to me the remains of fences which supposedly enclosed James Spencer's 'Excelsior Run' at the mountain summit; but no evidence of his possessing that run has been discovered in the records of the Lands Department. On 14 April 1862 he selected 40 acres at Waste Point (then called West Point). In a list of pastoral runs in the early 1880s, he or his son (also called James) is shown as the lessee of 15,000 acres in the 'triangle' at the foot of the range between Crackenback and Snowy Rivers. Within this triangle, James Spencer Junior also held a selection. Subject to further investigation, it looks as if the Spencer run at the summit is mythical.
[3] On Helms and his publications, see pp. 19–21, above.

143

just as much to find out how The Man from Snowy River was using the same land. Some of his findings were gloomy. In his *Report on the Grazing Leases* he wrote:

> A common, and, in my opinion, very improvident practice, will probably be continued as hitherto, viz., the constant burning of the forest and scrubs . . . This procedure has only a temporarily beneficial effect in regard to the improvement of the pasture by the springing up of young grass . . . I have seen some very detrimental effects from this practice here, because the heavy rains wash the soil away from the steep declivities . . . The more or less constant diminution of humus in the soil of the slopes is a danger not generally recognised.

In an address a few years later to the Royal Geographical Society of Australia, he returned to that theme. The graziers, he said, would be doing no damage to 'the dense carpet' of alpine pasture if they were not also fire raisers.

> Not satisfied with what nature yields, the herdsmen in order to improve the growth of the feed and make it sweeter, as they say, yearly burn large tracts of the grass and scrub. This procedure gives the otherwise fresh and cheerful-looking country here and there a desert-like appearance which is perhaps the least evil done. The greater evil is undoubtedly that it interferes with the regular absorption, retention and distribution of moisture . . . Where the natural growth covers the ground, numberless minute water-courses may be observed even in the height of summer and on considerably sloping ground, when at the same time the burned patches are void of all moisture . . . That ignorance and maybe greed should be allowed to interfere so drastically in the economy of nature is pernicious, and should not be tolerated. Even from an aesthetic point of view it ought not to be allowed, for what right has one section of the community to rob the other of the full enjoyment of an un-sullied alpine landscape, and to replace a fresh and fragrant growth by dead and half-burned sticks, making a desert of what was once a garden? The husbandman on the farm by the river, the artist and tourist who seek the picturesque, the botanist and zoologist who come in pursuit of plants and animals, are all interfered with. And why? Because some inconsiderate people are allowed to do as they please.[1]

Half a century later, the Lands Department was still allowing them to do as they pleased.

Nevertheless, the pressures for action were building up. In the early 1930s a distinguished forester, C. E. Lane Poole, launched a carefully planned operation of research into the erosion of soil, particularly the soil of mountain watersheds.[2] Economic depression cut that operation short,

[1] *JRGSA* vi (1896–8), esp. pp. 90, 91.
[2] Charles Edward Lane Poole (1885–1971) was born into a family which has made scholarly contributions to our knowledge of English and European history. He received his forestry education at Nancy, where Colbert's conservationist tradition of *Eaux et Forêts* was still

but not before its opening move, a *Report on the Murray River Catchment in New South Wales,* was carried to completion by a very determined forester, B. U. Byles. Like all the professionals of that generation, Byles still had a lot to learn – he learnt it later – about the under-story of grasses and herbs; but he knew the trees and shrubs, the soil on which they grew and the rock beneath the soil. Moreover, he was ready to live rough in rougher country than The Man from Snowy River ever penetrated. That man had no motive for forcing his way through the thick scrub of the steep slopes down to the bottom of the deep gorges; yet it was precisely on those steep slopes that the dangers of erosion were greatest. Byles' method was to ride along the stockmen's tracks with his packhorse on lead, to pitch camp in suitable places and then to examine the surrounding country on foot. In this way he spent six months exploring mile by mile the 700 square miles of Australia's most important catchment, and recording in his field books the findings of each day's exploration. Those field books were the unshakeable foundation of the report.[1]

Byles identified nine separate zones, which he defined by the types of vegetation that had adapted themselves to the various conditions of altitude, rainfall, temperature, soil and underlying rock. We shall look briefly at his findings in three of these zones. Above the tree line (6,000 ft), where low woody shrubs gave cover to alpine grasses and herbs, the fire-sensitive species had been severely damaged and the soil eroded in some places down to the bare rock. Although the area of total destruction was not as yet very great, the destructive processes could be observed almost everywhere. Unless these processes were checked, Byles gave warning, the consequences would be catastrophic.

In the snow gum zone (4,500–6,000 ft) he found only one stand of trees untouched or nearly untouched by fire. In the country running north from the summit area, very many trees had been destroyed and succeeded by woody shrubs, sometimes with a growth of grass beneath them. What troubled Byles most was the drying out of this country.

> The most notable example of this is the block covered by Snow Lease No. 29/57 ... I am told by men who worked on this block over 30 years ago, that their first job was to burn and keep on burning the woody shrubs and snow gum; at that time it required a very experienced horse to cross the swamps; once a horse put his hoof off a tussock of grass it would sink up to its belly in the swamp; now in an average summer a bullock dray can be taken across the former swamps and not sink more than a few inches ...

vigorously alive. He held positions of responsibility in West Africa and South Africa, before becoming Conservator of Forests in Western Australia and subsequently the Commonwealth's Inspector-General of Forests, Principal of the Australian Forestry School and head of the Commonwealth Forestry Bureau.

1 The report, dated 23 May 1932, appeared as Bulletin No. 13 of the Commonwealth Forestry Bureau.

Not only on this block, but throughout the Murray Plateau, the country is, on the testimony of men who have mustered cattle there all their lives, definitely drier now than it was 30 years ago. They point out again and again swamps and creeks which were formerly impassable but where now a man can ride without any danger of sinking. Consequent upon the drying of the swamps, the creeks are getting lower, and I can foresee the time when some of them will not last through the summer months.

Byles did not rule out the possibility of the drying up of this top country having been caused in part by a drier climatic cycle. However, rainfall records in the lower country – in the mountains there were none – did not support that explanation. Whatever the facts of the climatic cycle might prove to be, the destruction by burning of soil cover on the catchments was an established fact. Byles believed that he was safe in saying that it had had a very great effect on the steady drying process.

In the alpine ash zone (approximately 3,000–4,000 ft in the high rainfall of the western and south-western slopes) fires recurring at intervals of five to ten years had devastated steeply sloping areas where the erosion risk was extreme. Byles saw many scoured gullies up to 30 yards wide and 20 feet deep; they were bound, he said, to extend further up the mountain side with each winter's rain. He reported a landslide close to Geehi which had carried 5 acres of earth into the river, damming it back with a wall of earth 20 feet high. At Khancoban, 15 miles downstream, the river was thick with mud for two weeks following the landslide.

Summing up, Byles rejected despair. On a recent visit to the Taurus mountains in Turkey he had seen damage which had gone beyond the possibility of repair: damage to the Australian catchments, by comparison, was still in the early stages. Nevertheless it was bound to increase by geometrical progression unless prompt and resolute action was taken to prevent it. How to prevent it was a hard question to answer, because so little was known about the effects of grazing. The graziers themselves knew only one thing, that some pastures produced fat stock and others produced stock not so fat. The scientists did not know even that, because they had not as yet even begun to study the problems scientifically. Byles heavily emphasised the need for scientific research; but he was well aware that many years must elapse before its findings could become a clear guide to action. With damage to the catchments growing worse with every year, there had to be action now. Imperfect though his knowledge was, he believed that he knew enough to submit recommendations for action.

Some people wanted to stop grazing altogether, but Byles advanced two reasons for rejecting this policy: in the first place, fire fighters would no longer have easy access to the high country when fires got into it from lower down: in the second place, summer grazing in the mountains was a very great asset to two important primary industries, wool and meat. In addition to that, Byles did not believe that grazing by itself was doing any serious damage to the watersheds: fires were doing the damage. Yet he had made it quite clear on page after page of his report that in ninety cases out of a hundred these fires were started by graziers. He had explained their motives for starting them – to dry out the bogs where their horses and livestock got stuck; to destroy at the end of every summer the carry-over of woody, unpalatable herbage in the hope of getting a 'green pick' next spring. Knowing the graziers and stockmen as he did, he could have no doubt at all that they would never of their own free will change their ways. Consequently, he concluded his report with recommendations for the imposition of firm governmental control – rangers on patrol, gangs of men standing by to fight fires, better tracks to speed mobilisation at the danger points, equipment and seed for the rehabilitation of the eroded areas.

Money would be needed to bring these recommendations into effect. Byles suggested, for a start, that the annual income from all the mountain leases – £2,000 at that time – should no longer be paid into consolidated revenue but be spent within the area which produced it. He made no attempt at all to calculate the total cost. Had he done so, he would have realised that it was immensely in excess of the sum which any government of that time would propose, or any parliament approve.

PART IV

THE TWO LANDSCAPES

1. The tableland

During a scientific congress held a few years ago at Christchurch in New Zealand, two visiting botanists took a Sunday morning stroll in the pastoral country close by. They found a landscape entirely man-made. One botanist made a bet with the other that they would not see a single blade of native grass. He won his bet.

He would not have won it had he and his friend been taking their stroll in the pastoral country around Cooma. There, the pace of change has been slower. Even so, its direction is the same. Throughout Monaro, as everywhere else in the so-called high-rainfall zone of temperate Australia, scientific pasture management is creating a landscape in which exotic plants progressively supersede native plants. The explanation, in terms of natural science, has been stated as follows:

> All the evidence indicates that our native plants have neither actual nor potential value as artificially sown species. Though they may continue indefinitely to sustain our livestock in regions where artificial seeding is uneconomic, they will be progressively replaced in more favoured areas by plants from other parts of the world. This is because they suffer a serious disability – that they are incapable of high production, of response to high levels of fertility. They are adapted to the environment prevalent in Australia in the millenia before European settlement, to poor soils, to light grazing by nomadic, soft-footed marsupials and possibly also to drier climatic conditions than prevail today.[1]

Discovering those facts of life has taken much time and effort. 'Discovering', let us remind ourselves, is a continuing process, which begins with the observation of objects on the surface of the land and proceeds by probing the deeper levels.[2] In Australia, the opening phase of pastoral discovery lasted for nearly a century, with alternating moods of delight and disappointment – the squatter's delight when he found 'better country farther out', his disappointment when he found it so vulnerable to exploitation. A few squatters conceived the idea of improving the land by sowing it with grasses that had been familiar to them in England; but

[1] C. M. Donald, 'Temperate Pasture Species', in *Australian Grasslands*, ed. R. Milton Moore (A.N.U. Press, 1970), p. 303. In my amateurish explorations of the literature of this subject I have been particularly in debt to articles in the learned journals and in published symposia by C. M. Donald, C. H. Williams and F. H. M. Morley.
[2] See p. 12, above.

they achieved little success. The clover, burnet and rye grass which W. H. Hovell planted at every stage of his long ride to Port Phillip did not establish itself. He put the blame for this failure on the fire-raising blacks, not on nutrient deficiencies of the soil. William Farrar, by contrast, envisaged a programme of research into the chemistry of plants and soil.[1] In the late 1880s, that programme came to life in South Australia through the work of a Scottish scientist, William Lowrie, and an English nurseryman, Amos Howard. Lowrie saved the wheat farmers of South Australia by identifying a deficiency of phosphorus as the main cause of their steeply declining harvest yields. Howard identified and marketed a hitherto unnoticed legume, *trifolium subterraneum*, a stowaway immigrant of Mediterranean origin which had found an antipodean home on Mount Barker. In combination, Lowrie's superphosphate and Howard's subterranean clover started a pastoral revolution. Nodulation of the legume as a cheap source of nitrogen, the consequential improvement of soil fertility, the establishment of new legumes and new grasses, the dramatic climb of productivity per acre and per man – that is the rhythm of this still-continuing revolution.

In retrospect, it would be easy to exaggerate the speed with which the revolution gathered momentum. Howard made his first commercial sale of subterranean clover seed in 1906; but not until 1925 did the Prells sow it on Gundowringa station in the Crookwell district, and the Jeffreys sow it on Delegate station in Monaro.[2] Possibly they found a few hesitant followers; but two decades or more went by before 'sub-and-super' became the common practice of improving pastoralists. The explanation of this time lag between proven knowledge and practical action must be sought in the economic sphere. The man on the land has his living to make; but he fails to make it when he is too tightly squeezed between his costs of production and the price paid to him for his products. The great squeeze of the 1890s had ruined many thousands of graziers in many parts of Australia; but throughout the ensuing quarter of a century the cost-price relationship grew progressively favourable. That would have been a propitious time for heavy investments in superphosphate, seed drilling machinery and the other requisites of pasture improvement – if only the new knowledge had been sufficiently diffused; but by the time sub-and-super reached Gundowringa and Delegate economic depression was once again imminent. From the late 1920s onwards, sheer survival became of necessity the sole objective of graziers in Monaro, as elsewhere. Still, they climbed out of the depression and by the late 1930s were well poised for a massive effort of investment; but then came the war, with its physical

[1] See pp. 73 and 119, above.
[2] The Jeffreys had inside information on the experiment on Gundowringa since they were related with the Prells by marriage.

shortages. Thus it happened that pasture improvements did not get properly under way until 1947 or thereabouts. For the next four or five years it boomed, along with the booming wool prices. Even when the prices fell from their Korean peak, the investment continued on a massive scale, year after year. For a considerable time, its consequence was a strengthening of productivity which more than made good the weakening of wool on world markets. Not until the 1960s did drought and the cost-price squeeze begin to threaten the prosperity that had accrued from the pastoral revolution. As the decade approached its end, many people came to the conclusion that the pastoral industry was face to face once again with a struggle for survival.[1]

What has been written so far is necessary for background but has little specific reference to Monaro. Rather more relevant is *A Review of Production and Management on Southern Tableland Woolgrowing Properties*, published in May 1968 by the Bureau of Agricultural Economics. This study was based on a survey of 365 properties, approximately 15 per cent of the total number on the tableland.[2] The purpose of the survey was to measure productive efficiency in terms not of finance, but of physical resources and their use. The indices employed in varying contexts were wool per sheep, sheep per acre and wool per acre.

Prominent among the Bureau's findings were the following:

1. Deviations from average production – approximately 16 pounds of wool per acre – are very great. Some properties are producing more then 50 pounds per acre, others less than 5 pounds; more than one-third of the total are producing more than 20 pounds, more than one-third less than 10 pounds.

2. The higher yields come from improved pastures; the lower yields from pastures that are unimproved, or very little improved.

3. No clear correlation can be established between the size of properties and the volume of production per acre.

4. It is scientifically possible still further to increase the yield from improved pastures; but prudential calculations of various kinds restrain improving proprietors from exploiting these possibilities to the full.

5. The non-improving proprietors (so it is suggested in a single sentence on p. 29) are stupid people.

The *Review*, notwithstanding its painstaking statistical calculations, is in

---

[1] As early as 1962, F. H. Gruen issued a warning: see his 'Australian Agriculture and the Cost-Price Squeeze' in *Aust. Journ. Agric. Econ.* VI (1962), 1–20. The force of the warning was perhaps lessened by a temporary upswing of wool prices (on average 13.5 per cent) in 1963–4. Since then, the fall has been continuous; in 1969 and 1970 it became a nose dive. The figures are published at regular intervals by the Bureau of Agricultural Economics. The opinions of economists can best be followed in *Aust. Journ. Agric. Econ.* and *Quart. Rev. Agric. Econ.*

[2] Properties carrying less than 200 sheep were excluded.

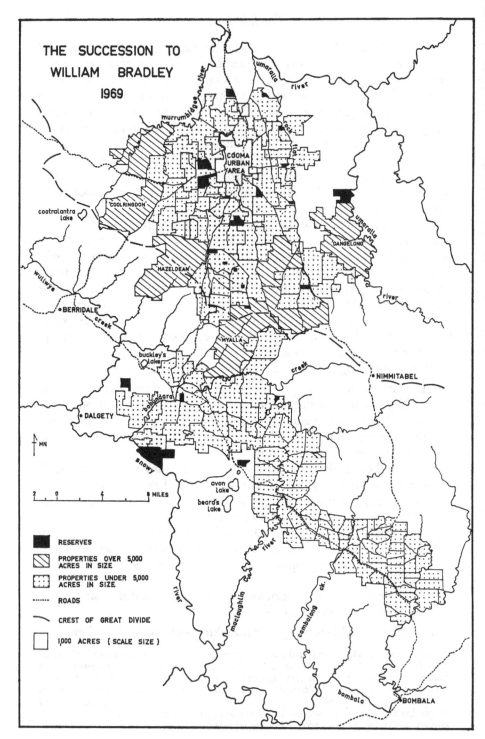

THE SUCCESSION TO
WILLIAM BRADLEY
1969

COOMA URBAN AREA

cootralantra lake

'COOLRINGDON'

'HAZELDEAN'

'DANGELONG'

umaralla river

murrumbidgee river

umaralla river

Wullwye

BERRIDALE

creek

'MYALLA'

buckley's lake

DALGETY

'BOBUNDARA'

snowy

NIMMITABEL

avon lake

beard's lake

MN

2   0   4   8 MILES

RESERVES

PROPERTIES OVER 5,000
ACRES IN SIZE

PROPERTIES UNDER 5,000
ACRES IN SIZE

ROADS

CREST OF GREAT DIVIDE

1,000 ACRES (SCALE SIZE)

river

maclaughlin

cambalong ck.

bombala

BOMBALA

Map 14

some respects remote from the facts of life in pastoral Monaro. In size, Monaro is barely one half of the tableland and is in many respects – altitude, temperature, soils, rainfall, duration of the dry season – conspicuously separate from the areas further north. Moreover, Monaro by itself (see map 5, p. 13) possesses almost as many climatic zones as those that the *Review* identifies on the entire tableland; certainly, it possesses twice as many zones as the three that the Bureau's fact collectors have made their sample. Finer brushwork is needed to portray Monaro's environmental diversity and the economic-historical consequences of that diversity. History, one well understands, is not prominently on the agenda of the Bureau of Agricultural Economics; but booms and slumps, droughts and floods have always been and still remain the urgent concern of wool growers, in Monaro as everywhere else in Australia.[1]

For these and other reasons it becomes necessary once again to explore the specifically historical evidences. A promise made earlier – to complete the sample study of land ownership on William Bradley's 300,000 odd acres – is now fulfilled in map 14. This map needs to be compared with its predecessors (pp. 93, 104, 105) and with the tables on pp. 89 and 103. From these comparisons the following completed table emerges:

Year	Over 5,000 acres	Less than 5,000 acres	Total
1884	7	76	83
1914	14	38	52
1969	5	151	156

Alas for the dreams of Henry Tollemache Edwards: 64 of the smaller properties on the map of 1969 are bits and pieces of the once mighty Bibbenluke estate.

At first sight, pasture improvement might seem to be a sufficient explanation of this fragmentation; a grazier who runs three or four sheep to the acre where he had been able before to run only one may well find it profitable to sell for cash two-thirds or three-quarters of his acreage. Actually, pasture improvement does to a considerable degree explain why soldier settlers of the 1940s have achieved better success, by and large, than their predecessors of the 1920s did. Nevertheless, historical chronology rules out the notion of a casual flow from pasture improvement to closer settlement and soldier settlement. As we have already seen, the closer settlement legislation is a product of the early 1900s, a generation before 'sub and super' made their first appearance in Monaro.

The intention of the closer settlement policy, like that of the free selection policy, was to create many farms where there had been before only a few sheep walks. The first Closer Settlement Act (1901) proved

[1] A discriminating agricultural economist has argued that variability of economic and environmental circumstance is *the* main problem of life on the land. See Keith O. Campbell, 'The Challenge of Production Instability in Australian Agriculture', *Aust. Journ. Agric. Econ.* 11 (1958), 1–23.

ineffective, because it upheld the principle of 'the willing vendor'; the second Act (1904) proved effective, because it asserted the principle of compulsory acquisition by the state. In each amending Act, as well as in the Soldier Settlement Acts which followed the two world wars, compulsion remained the rule. Quite often the mere threat of it proved sufficient: large proprietors sold many acres privately, in the hope of getting better prices than the state was likely to offer them. That happened more than once on Bibbenluke. There, the year of decision was 1925, when 42,000 acres were alienated. On Burnima, where H. T. Edwards had built his mansion, the years of decision were 1921–2 and 1945.

Our map demonstrates the real effects of a policy which was very persistently pursued; but it may possibly convey the false impression that legislation has put a stop to the ebb and flow of possession on the land. That, to be sure, was the intention of the legislators half a century ago. In their view, every farm ought to be a 'home maintenance area'. Unfortunately, they and their advisers were quite unable to define that area realistically: in the early years of closer settlement 750 or 500 acres or even fewer seemed to them sufficient for a farm in Monaro. Similar underestimations were made during the first attempt at soldier settlement. After the Second World War, the figuring became more realistic; but the concept of home maintenance still persisted. This concept is incongruous with the seasonal and economic variability on which stress was laid earlier. When 'average' rainfall and prices are as much the exception as the rule, the farmer requires not merely maintenance for himself and his family but assets which will carry him through the bad times. As one goes through the closer- and soldier-settlement districts of Monaro, one sees many signs that settlers who accumulated assets have bought out neighbours who failed to accumulate them. The causes of success or failure would appear to be various – quantity and quality of acres, initial capital and experience, ability and industry or their opposites, stability or instability of family life, good or bad health, good or bad luck.

Some day, let us hope, a pertinacious historian will cope with these and all the other economic and human complexities of land ownership and will clarify the patterns of change. Our concern, meanwhile, is with land use, as exemplified by the practice of particular users. It emerged from the *Review* which was discussed a few pages back that pasture improvement is conspicuous by its absence on a large number of properties on the southern tableland. For the much smaller area of Monaro the percentage figure of poor performance is almost certainly lower; but the inquirer who uses his feet and his eyes will assuredly see some large acreages of unimproved pasture. He need not for that reason endorse the suggestion, made in the *Review*, that non-improvers are stupid people. From the point of view of individualistic calculation, their reasoning may be shrewd: why should they face the risks and the trouble of a heavy investment in fertiliser, seed and machinery, when they already possess a sufficient acreage of unimproved pasture to give them a comfortable living? A farmer in Buganda, when urged by an agricultural inspector to plant more cotton, asked that question – 'Why?', 'So that you can buy yourself a motor car,' the inspector replied. The farmer sowed the cotton, harvested it, sold it at a good price and bought the car. Next year the inspector urged him to plant still more cotton. Again he wanted to know the reason why. 'When you have sold your crop,' the inspector told him, you'll be able to retire.' 'But I have retired already,' the farmer answered. In Monaro, one sometimes meets landowners who have retired already on their broad acres of unimproved or semi-improved pastures. Quite often they are interesting people. We, however, in the scanty space remaining to us, must confine ourselves to a few illustrative studies of their opposites.

Nowadays, the improvers of pastoral plants and soil are well served by scientific experts in the Department of Agriculture and in C.S.I.R.O. At Ginninderra near Canberra, C.S.I.R.O. maintains an experimental station which takes the southern tableland as its province. Not all its work has relevance for Monaro: for example, it envisages large extensions of commercial cropping on soil that has been enriched by the new pastures; but in Monaro the impediments to commercial cropping are formidable.[1] Nevertheless, Monaro men are deeply in debt to the scientists of Ginninderra for their work on pasture plants, nutrient-deficiencies of the soil, methods of controlling the weeds which compete with useful plants for the soil nutrients, methods of maximising the supplies of water, methods of sowing pasture seed, designs for improved machinery with which to sow it. Of course, it would be quite wrong to picture the improving landowner as a studious pupil, learning his lessons in the Ginninderra school; as in the past, he learns them in the practical business of earning his living. Nowadays, however, his practice is unselfconsciously in tune with scientific theory, and if he runs into an unforeseen difficulty – for example, a disappointing response of his pasture to dressings of superphosphate – he will more often than not take counsel with the people at Ginninderra.

We shall briefly consider five exemplars of pastoral improvement.[2] The

[1] See pp. 120–1, above. To the impediments there referred to, one should add the ubiquitous rocky outcrops which in Monaro inhibit cultivation except in scattered pockets. By contrast, the Ginninderra country is relatively free from this inhibition.
[2] What follows is based in large measure on replies received to a long and carefully constructed questionnaire, and upon a great deal of supplementary information, given in writing or in conversation. A by-product of this method of inquiry has been the location of rich documentary materials.

To my regret, I am able in this book to use only a small sample of the evidence which has been so generously put at my disposal. I hope that further use of it will be made in the future – by me or by some other historian.

first of them, Hazeldean, was a compact property of 36,000 acres in the 1890s; but it was cut down to a third of that size when James Litchfield divided his land (including Matong, a property of 8,500 acres south of the Snowy) among his four sons. Today, Hazeldean (11,525 acres) is integrated with two other Monaro properties, Biggam (5,600 acres) and Myalla (13,000 acres). James Litchfield's grandson, James Francis, has acquired these additional properties because he has had difficult problems to solve. When he took control of Hazeldean in 1923, he maintained its well-tested pattern of management – stud, breeding flock, wethers – but found himself in trouble with degenerating pasture, in particular, with corkscrew grass. In 1927 he made his first experiments with superphosphate; but the pasture did not respond. The previous year he had bought Biggam, a property with a good rainfall at and above the 3,500 ft contour line. Biggam was overgrown with green timber and overrun with rabbits; but it provided a safe refuge for the young sheep of Hazeldean when the corkscrew grass was dropping its seed. The tasks of improving both Biggam and Hazeldean were formidable – in the years of economic depression, almost hopeless; but from about 1934 onwards integrated management of Hazeldean, Biggam and the snow leases achieved impressive increases of production. The annual supply of good quality wethers then became the main problem; as a solution to it, Litchfield bought land to the west of the mountains. In 1967 he sold this land and bought Myalla – a good exchange, for his western property was in rapidly expanding wheat country, whereas he was a pasture man; besides, Myalla had been his grandfather's first home in Monaro and could be easily integrated with Hazeldean. Pasture improvement on Hazeldean had been resumed after the war with the same disappointing response to superphosphate as before, until C.S.I.R.O. was called in during the early 1950s; a deficiency of sulphur was then diagnosed as the cause of trouble and gypsum prescribed as the remedy. Since then, the main problem of management on Hazeldean has been how best to use the rich sward of clover, lucerne, phalaris and the rest. In brief, the answer has been to raise the production of the stud fourfold and to maintain the breeding flock and the wethers, reinforced by a herd of stud and commercial breeding Angus cattle. Thus pivoted, improvement of all kinds – paddocks, piped water, forage crops, machinery, buildings – continues year by year on Hazeldean.[1]

We shall review the other properties of our small sample in the descending order of their size. Delegate, just north of the Victorian border, was established in the earliest squatting years as an outstation of Robert Campbell's property – now the Royal Military College of Duntroon – on

---

[1] J. F. Litchfield finds improvement interesting. He is trying his hand at it not only in Monaro, but also on a large leasehold property in the so-called Ninety Mile Desert of South Australia – country which makes a good response to small dressings of trace elements.

the Limestone Plains. Campbell's unmarried daughter, Sophia, inherited Delegate; her initials S.C. are still the station brand. Since the death of Sophia quite late in the nineteenth century, the station has been continuously owned by members of the Jeffreys family, who are descendents of Sarah Campbell, Sophia's younger sister. Soldier settlement has chiselled this inheritance appreciably; but it still retains approximately 11,000 acres, almost all of which have been improved by superphosphate and the seeding of white clover, phalaris and rye grass. Comparative immunity from drought is ensured by the Delegate River and by a rainfall which on annual average – 27 in. – is well above the 19 in. of Hazeldean. Good management on many fronts – not least the rabbit front – has raised carrying capacity from 1½ dry sheep equivalent per acre in 1914 to approximately twice that figure today. Delegate's most troublesome problem is the acidity of its loamy soils. Excessive dressings of superphosphate, it would appear, have been making the acidity worse.[1] An experimental search for appropriate remedies is now proceeding.

Approximately 40 miles northwards of Delegate lies Bukalong. As we have already seen, this station holds a place of honour in the history of meteorological record.[2] Its present area – 2,550 acres – is much smaller than it was in 1884, when Garnock Brothers bought it from John Boucher; not closer settlement, but the division of Garnock land among numerous children, has been the cause of this reduction. Today, C. T. Garnock works Bukalong efficiently with the assistance, when he needs it, of seasonal and contract labour. He has 78 paddocks to look after, but they are all well supplied with dams or piped water. During the past twenty years he has planted nearly 10,000 trees to give his livestock winter and summer shelter. All his acres are improved with superphosphate and subterranean clover, phalaris, lucerne, white clover, cocksfoot and perennial ryegrass. Whereas in 1940 Bukalong was carrying barely 1 sheep to the acre, today it carries 3 to the acre – Merinos, whose yield of medium wool is approximately 12 lb per head. In addition, Bukalong carries a sizeable herd of cattle – Galloways, Herefords and some conspicuously successful crosses of the two. The Garnocks, like the Litchfields and all other good managers, are closely attentive to the market signals and would consider it foolish to put all their productive weight on one foot.

Mount Cooper lies 6 or 7 miles west of the main road, halfway between Nimmitabel and Bibbenluke. Like Delegate, it was an early Campbell squattage which the Jeffreys family inherited; but it passed later on into neglectful ownership and was compulsorily acquired in the 1940s for soldier settlement. The largest of the four farms into which it was divided

---

[1] How this can happen has been explained by C. M. Donald and C. H. Williams in articles in *Aust. Journ. Agric. Res.* v (1954), 664–87, and vii (1957), 179–89.
[2] See p. 81, above.

was 1,500 acres, the smallest 1,046 acres; but the production record of twenty years and more disproves the common assertation – well grounded, perhaps, before science enlarged all opportunities – that nobody can thrive in pastoral Monaro unless he possesses 2,000 acres at least. Of decisive importance at the outset were the characters and capacities of the soldier settlers; the father of one had been the manager of Bibbenluke, the father of another the manager of Burnima; all four were Monaro men with experience of life on the land. All four became close comrades; mutual love and mutual aid knitted their families together. Had it been otherwise, they could hardly have coped with the tasks which confronted them. From the windows of the old homestead – a beautiful house today, but a ruin in 1948, when the Sautelles took possession of it – one can see a hedged enclosure which serves no apparent purpose. Twenty years ago it was a corral into which the rabbits were driven for slaughter. Mount Cooper was then a moving grey blanket of rabbits; two soldier-settlers, in their first two years of residence, sold 18,000 skins – quite a useful source of cash income to make a start with.[1] When the rabbits were thinned out, the work of pasture improvement began; but it soon ran into trouble. The cause of the trouble, as on Hazeldean, was a deficiency of sulphur; the cure, as on Hazeldean, was co-operation between the practical man and the scientist. On the foundation of this cooperation, a firm and mutually rewarding friendship was built between John Sautelle of Mount Cooper and Peter Hutchins of Ginninderra. By the early 1960s, the Sautelle property was carrying 2,100 breeding ewes and 80 breeding cows – the equivalent of $3\frac{1}{2}$ dry sheep to the acre. The property was divided into 20 paddocks with a sufficiency of water in each; it was well supplied with shelter for the animals; it was in all respects a profitable and seemly property. In 1948 it had gone cheap because it was derelict; in 1962 Sautelle got a good price for it. He sold it because illness compelled him to look for a smaller property which would tax his strength less severely.

He bought Black Lake, the fifth and last property on our list. It lies about 4 miles distant from the tiny and charming township of Bibbenluke. Its area is 362 acres, of which approximately 80 are under water that is highly attractive to fly fishermen. The property merits inclusion in this survey as a visible refutation of a half-truth too often heard nowadays, 'Get big or get out.' Get capital, get experience – that would be better advice to give the prospective purchaser of a property, be it large or small. Sautelle bought the 362 acres of Black Lake at two-thirds the price per acre that he had received for his 1,046 acres at Mount Cooper: in consequence, his stock of capital, like his stock of experience, was sufficient. His new

---

[1] The rabbits were helpful in another way. The soldier-settlers bought their land cheap. In the 1950s, when their efforts, immensely aided by myxomatosis, cleaned the land of rabbits, the value of their land soared.

property was not seriously dilapidated; but its rabbits, red legged earth mites, thistles, briars and Bathurst burrs were a challenge to him in the early years. Its pasture was unimproved and its clay and alluvial soils were deficient in phosphorus, sulphur and molybdenum. Sautelle took counsel with his friend Hutchins and used aircraft to apply the appropriate fertiliser. He made many small paddocks and piped water to them. He grew fodder crops. While he was thus improving the carrying capacity of Black Lake – a conservative reckoning today would be 4 sheep to the acre – a fishing lodge managed by his wife produced a useful supplement to the family income. Meanwhile, he was building up a stud of Border Leicesters and winning prizes for his exhibits both locally and in Sydney. As his income from the Merinos fell, his income from the stud Border Leicesters rose. Thus did he prove that a small landowner need not succumb to the years of drought and the cost-price squeeze.

It will be timely now to look for common values in these highly diverse achievements of improvers, both large and small. The production value requires no further comment, unless it be the warning that even a fivefold increase of output per acre does not solve all problems of the individual and the community. Luxuriant swards may cause bloat in cattle; the enrichment of soils may cause too large a seepage of nitrogen into the flowing streams. Increases of productivity bring not only the expected prosperity, but also some unexpected headaches. Thus there is no rest for the intelligent improver, be he a man on the land or a man in the laboratory.

The conservation value of pasture improvement deserves to be underlined. Not one of the properties which we have been considering possesses any erosion worth mentioning. One looks forward with some confidence to the day when good pasture – on the tableland, not, of course, in the high country – will be doing the whole job which is now entrusted to officers of the Soil Conservation Service in Monaro. The conservation of water raises some particularly intricate and, at times, controversial problems. Standing on a hill after a month of heavy rain, one may look down on a dam which is no more than half full. One ceases to feel surprise when one realises that the compacted clay immediately surrounding the dam has been providing nearly all the run-off; most of the rainfall has been held by the thick sward on which one is standing. For the landowner, this creates a new problem; but he solves it to his own satisfaction by boring for water and pumping it to a large tank on a hill, from where it is distributed in cheap but durable polythene pipes to troughs in the paddocks. The people at Ginninderra are just as much satisfied with this newly emergent order as the landowner is: water, they assert, does its best work for production where it falls. If this assertion is well founded, the nation may do well to think more critically about its programmes of investment in irrigation storages and channels. Yet may it not happen that the improved pastures of Monaro will absorb all the rain that falls on them and still leave as much as before for the irrigated farms downstreams? C. T. Garnock, a close observer of change on Bukalong, feels pretty sure that flood flows in the gullies and creeks are now 90 per cent less than they were thirty years ago; he feels just as sure that the flow in the main stream is steadier and more permanent than it used to be. His explanation is that the deeper penetration of the moisture has increased its sub-surface percolation. It is for the hydrologists and associated specialists to assess the validity of that explanation; one assumes that they are purposefully at work, making their measurements and testing their hypotheses. Even so, water is too important a subject to leave entirely to the experts.

Aesthetic value is, or should be, a matter of shared concern to the expert, the practical land user and the ordinary citizen. The first white settlers saw beauty in the tall kangaroo grass; but some of their descendants will die without having ever having seen a blade of it, except perhaps on a railway embankment. Not only the native grasses, but the native shrubs and trees are yielding place to the exotics; of the 10,000 trees which C. T. Garnock has planted on Bukalong, almost all belong to the family of Pinus Radiata. Garnock, however, has left clumps of the native eucalypts on ridges and in other favourable situations. On Knockalong – a soldier-settler property which belonged five generations ago to Tombong, the Whittakers squattage[1] – the Wrights have left chain-wide belts of eucalypt on a plan conformable with the key-line method of land use. These belts provide firewood for men, shelter for sheep and a habitat for birds. They are useful: moreover, as one sees them either at ground level, or looking down from a small aircraft, they are beautiful. One needs to remember that the isolated eucalypt is a painfully vulnerable tree; if eucalypts are to be preserved at all, they must be preserved in clumps or belts. Surely they are worth preserving? The man-made landscapes of Monaro are many and diverse; each possesses its individual beauty; but Monaro's children should not be bereft of all opportunity to study and enjoy the native beauty which their forefathers knew.

Finally, plain human value: it is eroded when the continuities between past, present and future are too frequently and too roughly broken. In the Riverina, people used to talk of a twenty years' turnover of land ownership: even today, the population – barring its German element – appears footloose. In Monaro, by contrast, people have always shown a strong inclination towards staying put. Not only on Bukalong, but also on South Bukalong, Kyleston, Kuringai and Kilbreckin, families of Garnocks are living today no more than a dozen miles distant from the hillside where George Garnock built his hut 136 years ago. Litchfields have ridden more loosely at their anchorage; but within the borders of

[1] See pp. 38-9, above.

Monaro nine families in descent from the original James Litchfield are still owning land and using it well. Ryries, Jardines, Haslingdens, Woodhouses, Roses, McPhies, Moulds, Montagues, Hains – these names are no more than the beginning of the long roll call of family continuity. Some families, of course, have moved on while others have moved in; Monaro has thus been spared the stagnation which ensues when there is too little movement. More significantly in the present context, Monaro has been spared the disintegration which too much movement causes. Monaro is not an aggregate of human atoms; it is a deeply rooted society. People in Monaro do not merely own their land, they belong to it. The settlers of Mount Cooper, in the midst of their heavy toil to make the land fruitful and beautiful for the use and enjoyment of themselves and their children, do not forget the generations of men who lived before them on the same land. They are making it their task to record the names of the dead and to build their broken tombstones into a wall of remembrance.

2. The high country

On 21 August 1952 Dr W. R. Browne, who had been in days gone by the pupil, colleague and friend of the great geologist Sir Edgeworth David, delivered the David Memorial Lecture in the University of Sydney.[1] He spoke about 'our Kosciusko Heritage' – a theme appropriate to the occasion, for David had been a lover of mountains and a pioneer of glaciation studies on the Kosciusko plateau. In the first part of his lecture, Browne spoke as a scientist; in the second part he spoke as a moralist. Our heritage, he said, is precious beyond all price; but we are destroying it.

> Let us, then, imagine ourselves returning to earth after, say, 200 or 300 years, to observe how things are faring with our Kosciusko plateau. This is what we see. There is no timber in sight; it has all been cut down or killed off by bush-fires and rotted away. There may still be greenness in some of the valleys, but no longer is it that of the bogs and fens; they have all been drained and destroyed and replaced by snow-grass flats ... Along the valley of the inter-mittently-flowing Snowy River we see curious structures – dams and tunnels and power-houses – belonging to the pre-atomic age, now outmoded and long since silted up and choked with debris, neglected, crumbling to ruin ... In the mud-buried cities of Jindabyne and Adaminaby the diligent archaeologist digs happily, and with loving care unearths relics of the quaint twentieth-century Jindabynian and Adaminabite culture. There is no longer a soil-erosion problem in the highlands: there is little soil left to erode. The glory that was Kosciusko is a barren, stony desert; the spoilation of our heritage – man's triumph over Nature – is complete.

We may call Browne a Jeremiah, provided we bear in mind the words of the Lord to that formidable prophet: 'They shall not prevail against thee: for I am with thee, saith the Lord, to deliver thee.'[2] Browne, like Jeremiah, was not a moaner, but a fighter. He did not prophesy inevitable and final doom, but summoned his people to repent, to amend their lives, and to save their heritage.

Conditions in the high country at the turn of the half-century were a good deal more complicated than Browne gave his audience to understand. A revision of the snow leases in 1944 had revealed increasing prudence within the Lands Department: for example, limits were set on each

[1] *Aust. Journ. Sci.*, Supplement, xv, no. 3, December 1952.
[2] Jeremiah 1, 19; 7, 3; 24, 6.

V   Kiandra's postman

VI   Kiandra Snow-Shoe Club

VII   The high country (chromolithograph by Eugene von Guérard), mid 1860s

VIII   Wragge's observatory on Mt Kosciusko, about 1900

block to the intensity of grazing, and rangers were appointed to see that those limits were respected. Unfortunately, the rangers were too few; moreover, nobody at that time knew enough about the mountain pastures and the grazing habits of sheep to specify realistic limits.[1] Another decision of the Lands Department in 1944 was almost as much a venture into the unknown. The big men from the far west were pushed out of the high country: henceforward the snow leases were to be reduced in size and reserved in the main for the smaller men who lived closer at hand.[2] As a symbol of the new deal, the administration of the leases was shifted from Wagga Wagga to Goulburn. What effects followed from these new arrangements would be hard to say. Smaller blocks, more carefully fenced and more regularly in use, may well have intensified the grazing pressure. The Litchfields, certainly, would have found it more difficult than in the past to ease the pressure by moving their animals up and down the mountain, slowly following the retreat of the winter snow and prudently anticipating its return. For them, however, that problem did not arise, because good management on their home properties was beginning to give them as much grass as they needed, at the very time when the costs of mountain grazing were rising. Not long after the end of the war, they retired from the snow-lease business. If other good managers were doing the same, it would seem to follow that the high country was becoming increasingly a refuge for the less efficient. The David Memorial lecturer, however, showed no interest in such fine distinctions; he condemned mountain grazing in all its sizes and shapes. On his own assumptions, he ought therefore to have acclaimed the decision taken in the mid-1940s to withdraw from occupation the very highest country. To the snow lessees, that decision proved to be the thin end of a much-hated wedge.

Still another decision taken at that time – to establish the Kosciusko State Park – might have given great comfort to Dr Browne, had not his prophetic purpose forbidden him to see light amidst the encircling gloom.

[1] Originally there were four rangers, but the number soon fell to two. Sheep in the high country graze selectively on the small plants between the tussocks of snow grass: consequently 1 sheep to the acre, which may appear to be a prudent stocking rate, represents a grazing pressure many times heavier.

[2] I owe the following table to Mr W. G. Bryant of the Soil Conservation Service.

Summarised Percentum Distribution of Home Addresses, Snow Lessees 1943–50

Approximate distance from lease:	East %	West %	Total %
Less than 25 miles	33.4	2.5	35.9
26 to 50 miles	25.6	8.3	33.9
51 to 100 miles	12.7	5.1	17.8
101 to 150 miles	1.3	2.6	3.9
Greater than 150 miles	—	1.2	1.2
Unclassified			(7.3)
	73.0	19.7	100.0

Contrariwise, two groups of zealots, the bush walkers and the crusaders for catchment protection, had welcomed the Kosciusko State Park Act, 1944. We must consider the origins of these two groups.

The bushwalkers were hardy young men from the cities who spent their holidays exploring the high country; all of them enjoyed living rough; some of them became sensitive students of rocks, soil, plants, animals and insects. They formed clubs and a federation of clubs, which soon became recruiting ground for a very determined conservationist, Myles J. Dunphy. In the early 1930s Dunphy called for volunteers to establish a National Parks and Primitive Areas Council. In 1938, this council drew up proposals for a Snowy-Indi Primitive Area – a million acres of mountainous country in New South Wales and Victoria. This proposal, strongly supported by other groups of bushwalkers and conservationists, aroused the interest of W. J. McKell, the wartime premier of New South Wales.

McKell was just as much a supporter of the movement to protect the water catchments. In the much longer history of that movement, a conspicuous landmark is the Royal Commission appointed in 1902 to investigate the conservation and distribution of the waters of the River Murray and its tributaries. A decade and more went by before action followed; but in 1915 the Commonwealth and the three riverine States instituted the River Murray Commission, which anticipated by nearly twenty years the Tennessee Valley Authority of the United States.[1] The Commission's first big work of water storage – the Hume dam, 10 miles upstream from Albury – was completed in 1936. About the same time, riverland people, who abhorred the prospect of big loads of silt being carried into the storages, formed the Murray Development League. This organisation, with its well-edited monthly *The Riverlander*, gave strong regional support to the cause of catchment protection. Academic people began to see the need for scientific study of the watersheds: in 1941, Miss S. G. M. Fawcett[2] was seconded from the Botany Department of the University of Melbourne to the Soil Conservation Service of Victoria, for the express purpose of studying soil cover in the high country and the changes imposed upon it by summer grazing. Her researches established new principles and techniques of ecological investigation in Australia. In the mid-1940s, she assisted Judge L. B. Stretton, who as Royal Commissioner was conducting an inquiry into the condition of the catchments. His eloquent report alerted the Victorian public to the need for immediate strong action in their defence.[3]

1 The R.M.C. is a planning, not a constructing authority. Construction of the works agreed upon is a State function.
2 Later Mrs Denis Carr.
3 *Report of the Royal Commission to enquire into Forest Grazing* (Govt. Printer, Melbourne, 1946). Judge Stretton's earlier report on fire prevention, following the disastrous bushfires of January 1939, had made a powerful impact on public opinion.

In New South Wales, ministers and senior officials of the Forestry and Agricultural Departments had been moving more quietly but no less purposefully along the same path. In 1933, they had established a Soil Erosion Committee; in 1938, this committee came of age as the Soil Conservation Service. Its official head, E. S. Clayton, had been the senior experimentalist in the Department of Agriculture and one of the judges in the annual wheat competition; but he saw little sense in striving for higher yields of wheat when all the time the soil was being blown or washed away. He saw even less sense in cutting up good catchment country to make starveling farms. That, precisely, was the ambition of a patriotic and flamboyant Monaro man, District-Surveyor C. J. Harnett.[1] Towards the end of the war, Harnett produced a plan for closer settlement of Snowy Plain, the broad, grassy mountain valley of the beautiful Gungarlin River. When Clayton saw that plan, he went to the Premier, W. J. McKell, and the Minister for Lands, J. M. Tully, and persuaded them to come riding with him for two weeks in the high country. It was on that ride, we may well believe, that the decision was taken forthwith to draft a Bill to establish the Kosciusko State Park.

Introducing the Bill on 22 March 1944, the Minister of Lands affirmed and expounded three principles: first, permanent preservation of all the water catchments within the boundaries of the Park; secondly, permanent reservation and development of the Park for the recreation and enjoyment of the people; thirdly, the controlled use of Park land for pastoral purposes, insofar as they were consistent with the first and second principles. Under no circumstances, the Minister declared, would the commercial activities of man be permitted to despoil the Park. The Bill passed through all its stages quickly and with general support. The Kosciusko State Park Act became operative by proclamation on 5 June 1944. The Act established a park of approximately 1,300,000 acres and entrusted the 'care, control and management' of that superb public estate to the Kosciusko State Park Trust.[2]

Even W. R. Browne must have allowed himself to do some wish-thinking when that Act was passed. It was as good an Act as anybody could have hoped for, except in two particulars: the provision which it made for the composition of the Trust, and the provision which it failed to make for the Trust's income. The composition of the Trust was as follows: the Minister of Lands as chairman, five nominated officials (among whom were included representatives of the forestry, soil conser-

[1] In the Harnett home in the Eucumbene valley, Barcroft Boake (see p. 137, above) had spent two unusually happy years.
[2] The external boundaries of the Park, as defined in the Schedule to the Act, enclosed a great variety of landscapes at altitudes rising from 1,000 to 7,314 ft. Not all the area was crown land; there were enclaves of freehold which the Trust was empowered to acquire by purchase. Some of these enclaves still remain (1970) unabsorbed.

vation and tourist services) and two persons, presumably laymen, to be nominated by the Minister of Lands. Before very long, the Lands Department interest, which still remained in large measure a revenue interest, established its predominance at meetings of the Trust. An amending Act of 1952, which empowered the Minister of Lands to nominate four non-official members instead of the original two, made that predominance decisive. Trustees who held strong conservationist convictions, most notably E. S. Clayton and B. U. Byles, found themselves repeatedly in a minority on issues which they considered vital.[1]

The Act laid it down that the Trust should administer its own income; but the only sources of income which it enumerated were: (1) rents, royalties, fees and fines; (2) such moneys as Parliament might provide. In the event, the Trust remained for many years almost entirely dependent upon the rents which accrued to it from the snow leases – an intolerable situation for men who might find themselves unable to perform the first duty laid upon them, protection of the mountain catchments, unless those catchments were closed to grazing. If Parliament had voted direct and substantial grants of money to establish and maintain the Park, it would have freed the Trust from that crippling contradiction; but Parliament and the public had fooled themselves into believing that the Park they *thought* they wanted could be got on the cheap.

That was the position in the early 1950s, when Browne delivered his David Memorial Lecture. He thought the Park worth four sad sentences. By contrast, he spent many sentences and much emotion in discussing and denouncing the Snowy Mountains Hydro-Electric Authority. That massive engineering enterprise (four years junior to the Park) was able to spend a million pounds for every thousand pounds which the Park Trust was spending. Its mighty works are too large a theme to be crammed into a single chapter of this book; its prehistory and early history have already been told, and told well.[2] Let us therefore confine our attention to the areas of contact, co-operation and conflict between the Authority and the Trust. Browne called the Authority a 'monster'; he called its works a 'desecration'. Those words were violent; but in fairness to Browne two things need to be said: first, the Americans, when they established the earliest national parks in the 1870s, had refused admittance to the hydro-electric and irrigation engineers: secondly, no engineer in any country has ever taken

[1] At this point I must acknowledge my great indebtedness to the Minister of Lands, Hon. T. L. Lewis, M.L.A., who has given me unrestricted access to the Park's records. I am also greatly indebted to Mr B. U. Byles, who has put at my disposal his remarkable private papers. In doing so, he did not realise that I should find them significant not only historically but also biographically. A third debt which I must acknowledge is to Mr Neville Gare, Superintendent of the Park from 1959 onwards, who has allowed me to read the unofficial Byles–Gare correspondence.

[2] Lionel Wigmore, *Struggle for the Snowy* (O.U.P., Melbourne, 1968).

rivers out of their beds and put them into concrete pipes without making eyesores.[1] Browne saw and hated new and raw eyesores in his beloved mountains; 'outrageous' was his word for them. He saw bulldozers gouging roads and channels athwart the steeply sloping mountains and in their tracks he saw erosion. 'Look your last,' he told himself, 'on all things lovely.'

Browne, however, did not see into the mind of Sir William Hudson, the man who bore responsibility for the construction of 17 major dams and many smaller ones, 9 power stations, 80 miles of aqueducts, 100 miles of tunnels and many hundreds of miles of roads, not to mention all the workshops, houses, restaurants and shops which served his work force. Hudson laid a heavy hand on the wilderness. Nevertheless, he hated erosion just as much as Browne did, and for a very good reason: erosion would send silt into his dams. It needed a man who could get to know Hudson by talking shop with him to understand this. That man was E. S. Clayton, the pioneer of soil conservation in New South Wales. His aim from the early 1950s onwards was a Clayton–Hudson alliance.

Clayton has put on record his carefully formulated conclusions on catchment policy in the Snowy Mountains.[2]

> It is essential to preserve, or indeed improve if possible, the vegetation cover to conserve snow, increase fog and cloud drip, improve penetration, reduce evaporation, improve continuity of flow and raise rather than lower the groundwater table. This requires a type of vegetation which allows the rain to enter the soil cover before reaching the stream. In this way erosion is kept to a minimum and flash flood run-off reduced. All informed authorities agree that any reduction of the ground cover in the Snowy watershed reduces water infiltration and catchment efficiency . . . and unless the deterioration is arrested the damage to the source of usable water would reach devastating proportions.

Clayton was well aware that the Authority's bulldozers had done a lot of damage to the catchments in the early years; but he was aware also of Hudson's determination to repair that damage so far as he was able and to take every possible precaution against further damage being done. To discover the techniques of repair and prevention and to teach those techniques to the engineers and labourers was bound to take time; but from one year to the next the Authority improved its conservationist performance.[3]

---

[1] See John Ise, *Our National Park Policy* (Johns Hopkins, 1961); also Fraser Darling and Noel D. Eichorn, *Man and Nature in National Parks* (Conservation Foundation, Washington D.C., 1967). Eyesores? The empty bed of the Snowy River below Island Bend will serve as an example.

[2] *Catchment Protection in the Snowy Mountains* (S.M.A., Cooma, 1967) p. 12.

[3] Clayton (*Catchment Protection*, pp. 19–25) describes the methods employed by the Soil Conservation Service and the Authority. The latter's expenditure on the restoration and protection of soil cover up to the mid-1960s was approximately £3,000,000.

The two landscapes

Meanwhile, as a participant in the Hume–Snowy Bush-fire Prevention Scheme (inaugurated in 1951) the Authority was massively reinforcing the fire brigades. At the same time it was reinforcing the scientists, who remained until the mid-1950s spread far too thin in the mountains. Engineers, it is often said, have no time for ecology; but Hudson had time for it. His team of four highly trained researchers made important quantitative studies of erosion (and the techniques for controlling it) of siltation, of soil cover and of the snow leases. That last study was both ecological and historical.

In Hudson's mind the conviction took shape that it was not enough for him to tidy up after his own bulldozers: summer grazing on the mountain catchments was the main threat to his water channels and storages – a threat that would increase by geometrical progression so long as the snow-lease system survived. Clayton held the same conviction. In the mid-1950s the two men launched a combined assault against grazing above the 4,500 ft contour line.

Within the corridors of power there was bitter in-fighting which went unreported in the newspapers; but two statements by non-governmental associations made front-page news. In April 1955 the Murray–Murrumbidgee Development Committee (an extended arm of the Murray Development League) published an interim report entitled, *The Condition and Administration of the Murray–Snowy–Murrumbidgee Catchment Area.* That report followed a tour of inspection which had been planned with great care by the moving spirits of the Development Committee, John Redrup and G. V. Lawrence. Invitations to join the tour had been issued to Surveyor-General G. W. Vincent, E. S. Clayton of the Soil Conservation Service and A. B. Costin, a former member of that Service, who had become well known by now for his ecological research in the high country.[1] Vincent and Costin accepted the invitation but Clayton excused himself; civil service anonymity best suited his book; besides, he knew that everything which he would want to say would be said by Costin. When the report appeared it bore the Clayton–Costin stamp. Its main recommendations were:

1. Elimination of the high altitude snow leases.
2. Massive reinforcement of the fire prevention services.

[1] A. B. Costin began his career in the Soil Conservation Service of N.S.W., specialising on high catchment field work. In 1950–2, as a post-graduate student of University of Sydney, he wrote a thesis which grew into an impressive book, *The Ecosystems of the Monaro Region of New South Wales* (N.S.W. Govt. Printer, 1954). Thereafter he spent some years with the Victorian Soil Conservation Service, doing fieldwork in the Alpine catchment area. Meanwhile, an American ecologist, L. Gay, had been conducting field research in the Kosciusko area with support from C.S.I.R.O. In 1957 Costin joined C.S.I.R.O. and was assigned the task of establishing a permanent Alpine research station. Work still continues there under the direction of D. Wimbush.

3. Catchment research at a C.S.I.R.O. field station to be established in the Snowy Mountains.
4. Additional provision for the conservation work of the Snowy Mountains Authority.
5. A new and powerful Murray and Snowy Mountains Catchment Authority.

Those recommendations were anathema to the Lands Department and the snow lessees; but they made a powerful impact upon public opinion.

In May 1957 the Academy of Science joined the battle. Its *Report on the Condition of the High Mountain Catchments of New South Wales and Victoria* was the work of a strong committee chaired by Professor J. S. Turner, the Melbourne botanist who in 1941 had sent the first ecological field worker into the high country.[1] The report contained a comprehensive survey, supported by an annotated bibliography, of the progress of knowledge from the time of Helms to that of Costin. In their discussion of the scientific evidence, Professor Turner and his colleagues were scholarly and cool in tone, but their conclusions and recommendations were nonetheless forceful. On all essential issues, the men of science stood shoulder to shoulder with the riverlanders against the snow lessees.

The struggle came to a head when the Catchment Areas Protection Board, meeting under Clayton's chairmanship, imposed a veto on the renewal of the snow leases above 4,500 ft. The immediate sequel to that bold act was a counter move by the Kosciusko State Park Trust, which claimed overriding jurisdiction on catchment management and every other issue arising within the park's boundaries. At that point in time, the officials and laymen who supported the snow lessees held a majority of one on the Trust. Thus it came about that the Government of New South Wales had to make a decision between a recommendation from the Catchment Areas Protection Board to terminate all leases above 4,500 ft, and a recommendation from the Kosciusko State Park Trust to renew those leases. A split in the cabinet appeared imminent. The Minister of Lands, Roger Nott, threatened resignation if the leases were not renewed; the Minister of Conservation, Ernest Wetherell, threatened resignation if they were not terminated. Cabinet decided to terminate the leases. Nobody resigned.[2]

---

[1] See p. 166, above.

[2] The Ministry of Conservation had brought under the same umbrella the hitherto separated services of forestry, soil conservation and water supply (including irrigation). Wetherell, as Minister, gave his full support to Clayton from the beginning to the end of the crisis. He had spent his impressionable years in Broken Hill, during the period when tree planting by a large mining company was beginning to rescue the citizens from the appalling dust storms that had been blowing in from the spoiled saltbush country to the west of the city. That experience made him a staunch conservationalist. One takes pleasure in pointing out the connection between the first dramatic victory of the conservation movement and the decision of a large capitalist undertaking, many years earlier, to spend some of its profits in planting trees.

The two landscapes

The termination of the high altitude leases deprived the Kosciusko Park Trust of the greater part of its income, but the Snowy Mountains Authority made the loss good by an annual grant of £15,000, to be spent on conservation work in the catchment area. The Trust delegated the spending of that money to the Soil Conservation Service – that is to say, to Clayton. He was finding it impossible to forgive the Trust for the part it had played in the recent crisis: except when items which concerned him officially were on the agenda, he no longer attended its meetings. He could see no point in them. Grazing in defiance of the recently imposed ban was rife in the park; but the Trust took no effective steps to stop it. It lacked the means to take any effective steps in any direction. In 1958, the fourteenth year of its hitherto inglorious life, it possessed the following establishment: in Sydney, a retired official of the Lands Department, acting as part-time secretary: at Waste Point, a caretaker, an electrician, a plant operator and two park workers.

From this pit of futility the Trust fought its way upwards. How it did we shall begin to see if we imagine ourselves on tour in the mountains with B. U. Byles. The record of his many tours survives in his field-books and in the journals which he wrote out by typewriter as soon as he returned home.[1] Choosing one of these journals at random, we find ourselves encamped with him one evening in late summer at Farm Ridge hut, well into the mountains south of the Eucumbene River. His pack, which weighs 34 lb, contains maps, field glasses, a field book, spare socks, a change of underwear, a small towel, a cake of soap, and rations for five days: that night, rats eat a corner of his cheese, a packet of his soup cubes and all his soap; but wood ash, he reminds himself, is an excellent substitute for soap. After breakfasting on Uncle Toby porridge, salami, dried fruit and coffee, he sets out along the trail that leads south to Grey Mare hut. He stops to look westwards to Jounama, and to recall his first ascent of that mountain half a century ago; then and there he had taken the decision to become a forester. Soon he starts walking again. From time to time he gets down on hands and knees to identify the small plants which are beginning to re-colonise the bare patches between the snowgrass tussocks. This reminds him of words that he had said not long ago to Sam Clayton – that if Clayton had done nothing else in his life but get the stock out of the mountains, he would be able in his old age to look back on a life well spent. By 5 p.m. he is at Grey Mare hut, starting to get everything shipshape for the three soil conservation people who are joining the tour. They arrive half dead with fatigue shortly after sundown.

The next three days are not too strenuous. The air at 6,000 ft is crisp

---

[1] His friends said that he suffered from typewriteritis. I am choosing the journal of a tour he made in March 1964, although that means taking a leap ahead in time, because I have happened recently to cover a good deal of the same country with a pack on my back.

and the sun warm. The party devotes a lot of time to a 'hands and knees' examination of soil cover in the high valley of the Valentine River. At lunch one day they formulate 'a conservation and vegetation restoration plan': every walker henceforth will place his 'daily offering' carefully in the middle of an inter-tussock space, sprinkle thereon some appropriate seed and mulch it with dead snowgrass. 'This conforms,' Byles observes, 'to the fundamental conservation philosophy of making good use of what is generally regarded as waste material.' Meanwhile, he is making good use of his eyes. On his way to the Valentine Falls he sees survey pegs inclining down hill and asks himself whether the angle of their inclination may not be a measure of the earth creep which has occurred since they were driven; he feels sure that he can reasonably put this question, because a surveyor's chainman invariably drives his pegs vertically. That afternoon, he and his party spend some hours inspecting the broad valley at the head of the Valentine River:

> To any intelligent questioning observer it must be apparent that something has happened to this valley. In locations such as this Nature just does not make creeks with sandy eroding banks, sandy beaches and beds of shingle and stones. Costin advances the theory that the acceleration run-off from the slopes, caused by destruction of the hillside bogs and the destruction of a significant amount of the herb-snowgrass cover, has caused a scarcely visible, meandering, slow-flowing, scrub-lined brook to become widened, deepened and accelerated; this has lowered the water table in the valley bottom, thus concentrating more water in the single creek and thereby accelerating the process. This theory certainly fits the observable facts . . . and is welcome to a prejudiced conservationist intent upon building up the vegetative cover of the catchment; but no-one can claim certainty when he attempts to use observable present day data in order to reconstruct the past.

During the next few days that thought recurs many times in his journal: the data we collect will never enable us to reconstruct the past with complete certainty. Nevertheless, he insists, we must keep on collecting data – ecological data, historical data, every kind of data which has a bearing on the management of the Kosciusko State Park.

On this tour, as on every other that he made, Byles devoted a great deal of thought to the principles and programme of his action as a trustee. By the time he returned home he would have material for various memoranda to be submitted to his colleagues at their next meeting: memoranda on practical details such as garbage disposal or the clearing of bush paths in the neighbourhood of Sawpit Creek: memoranda on large issues of policy, such as the 'primitive' area and – arising out of that – a proposal to divide the Park into zones. The Trust, he is well aware, has a responsibility towards the people who enjoy fun and games at Smiggin Holes, but it has also a responsibility to preserve the wilderness so that men of science may study it and lovers of solitude find healing in it. He affirms this dual responsibility as he bids farewell to the Valentine.

> It is a vertible walker's paradise ... one may bathe in the creeks and rivers, feel the beds of sphagnum, stroll around them and study the flowers: hike vigorously at top speed over hill and dale or climb to the top of every rocky knob. No ticks, no leeches, no snakes, no mosquitos. Possibly a friendly blowfly or a march fly or two but nothing venomous. But to the walker the joy of this area depends on its remoteness and the fact that some effort is required to get into and over it. Without the effort there would be but little joy. If it were made easily accessible by motor road the entire appeal for the walker would be lost. The Park is a large one; we must plan it for the enjoyment of all legitimate tastes. Those who like metropolitan comforts and the metropolitan scene may get both at Smiggins, in the Perisher and at Thredbo. Should they be allowed to spoil all the rest of the Park for those who want to get away from the metropolitan scene? This is part of the fundamental philosophy of National Park management.

Byles had a fellow feeling for 'tiger bushwalkers'; but for companionship on a tour he preferred hands-and-knees observers, particularly if they possessed what he called 'pothooks of the mind' on which to hang their observations. His own scientific training as a forester, reinforced as it had been by experience and study in later years, had given him a good supply of pothooks; but men of science, he insisted, did not possess a monopoly of them. He made many attempts to persuade lay members of the trust to go walking with him and had an enjoyable outing one day with Leo Barry, a grazier and political leader with deep and tough roots in the Snowy country. He felt sure that he and Barry had much to learn from each other, despite the fact that they had taken different sides in the battle of the snow leases. A walk which Byles had arranged with Sir Garfield Barwick in the aftermath of that battle produced long-continuing consequences. Barwick, too, had voted – as Byles judged the issue – on the wrong side; but he had promised to reconsider his position if Byles produced the additional evidence which he, as a lawyer, considered essential. Byles took him at his word. In January 1958 Barwick, Costin and Byles set out on a three days' mountain tour. By the time it was over, the conservationist minority on the Trust had become a conservationist majority.[1]

Things began to move fast in the Kosciusko State Park. The Trust advertised the post of Park Superintendent. A young forester, Neville Gare, felt all the more determined to get that post when Byles warned him that he would have to perform the labours of Hercules. In May 1959, at the start of the ski-ing season, Gare took charge of his staff of four. He

---

[1] Barwick at this time was about to enter Commonwealth politics. After holding office first as Attorney General and later as Minister of External Affairs, he became not only Chief Justice of Australia but also President of the Australian Conservation Foundation.

found himself working with the labourers to clear a track through the snow for the motor cars and buses; he found on his table a list compiled by Byles of fifty tricky tasks awaiting his immediate attention. At the same time, news reached him of discussions proceeding in Sydney with an entrepreneur of the tourist industry who wanted to build a chairlift up the south-western escarpment to the summit of Mt Townsend. Meanwhile, he had received notification that his attendance was not required at the forthcoming meeting of the Park Trust. 'I feel,' he told Byles, 'rather like the manager of a departmental store being kicked out of the directors' meeting.' Gare was not a master of the art of concealing his feelings. Explosions of protest occurred in his letters and memoranda to the trustees: 'I strongly object . . .', he told them, or 'I insist . . .'. That, said Byles, was the wrong tone for him to take: the Trust could give orders to the Superintendent and insist that they be carried out; but the Superintendent must never on any account entertain the idea that he could insist on the Trust obeying his orders. On one occasion Byles asked Gare to confess a mistake that he had made in a too-impulsive letter:

> It takes courage to lose face deliberately but faces are a bloody nuisance and we would all be a lot better without them. Let us save other people's faces at all costs but lose our own as soon as we possibly can. Then we are free to tell and act the truth.

He himself, Byles confessed, had spent a long time learning that lesson: as a young man, he too had wanted to blast a passage through the rocks and shoals; but his preference now was to navigate around them – whenever possible. It might not always be possible.

Gare's tactics of frontal assault and Byles's flanking tactics made a strong combination. Notwithstanding their temperamental differences, the two men remained unshakeably at one with each other in their service to the common cause. They exchanged drafts of policy documents which they were compiling at home after the day's work was done – a document on the rights and duties of the secretary and another on the Trust's financial needs; a plan for the further development of Perisher as a tourist resort and the outlines of a master plan for the whole Park. In due course, these documents were submitted to the trustees for discussion, criticism, and eventual approval. Action followed. Within two years Gare's staff rose from five to fifteen; he had three assistants in the office and two rangers on patrol in the mountains. The Trust still contained diversities of attitude and opinion; but it was no longer afflicted by paralysis of the will. Late, but not too late in the day, it began to take hold of the hitherto neglected powers belonging to it by Act of Parliament.

Section 5 (iii) of the Act of 1944 declared:

> The Trust may retain as a primitive area such part of the Kosciusko State Park (not exceeding one tenth of that Park) as it may think fit.

In 1945, the bushwalkers approached the Trust with a plan for giving effect to that enactment; in 1946 the Royal Zoological Society and the Linnaean Society of New South Wales made a combined approach. To be sure, the men of science and the bushwalkers held different conceptions of the primitive area. The Trust appeared in those early years to hold no conception of its own; at any rate, it took no action. By contrast, the Snowy Mountains Authority, from the time of its foundation in 1949, started purposefully to make plans for engineering works in what was bound to be the heartland of the primitive area – if, and when, and in whatever way, it should be defined. Those plans were not secret. The Authority explained and illustrated them in an information brochure which was available from the early 1950s onwards to every interested visitor to the Snowy Mountains.

In the late 1950s, men of science began to show a revived interest in the idea of a primitive area. The Academy's report of 1957 on alpine grazing had referred in passing to the idea and had suggested that the Authority should re-examine its plans for a dam on Spencer's Creek below Mt Kosciusko. In 1958, a large and influential group of scientists from Sydney and Canberra expressed alarm at the Kosciusko prospect and the closely related project of an aqueduct down Lady Northcote Canyon and on to the Geehi Dam. The Minister of Development, Senator Spooner, told the scientists that their protest had come too late, because engineering work had already been started in the disputed areas. In disregard of that official notice to keep off the grass, the Academy of Science published in 1961 a document entitled *The Future of the Kosciusko Summit Area: A Report on a Proposed Primitive Area in the Kosciusko State Park*. The Academy did not reiterate the objections, which were predominantly scenic, to the works at the Geehi end; but it expressed firm opposition to some of the works at the Kosciusko end. Without those works, the Authority maintained, the entire project would be crippled.

On 10 January 1963 the newspapers published a long statement by Mr Howard Stanley, the deputy chairman of the Kosciusko State Park Trust. The Trust, he said, had decided to set aside as a primitive area 70 square miles of the Park.[1] Action was urgently required 'to protect for the nation a magnificent scenic area and a living museum containing many of

---

[1] A survey subsequently undertaken revealed that the area was 97 square miles. In all essentials, it was identical with the primitive area recommended by the Academy. Consequently, it imposed no veto on the Authority's proposed Kosciusko dam; but it did impose a veto on water-gathering aqueducts which the Authority considered essential if the dam was to fulfil its function.

the nation's natural masterpieces'. The primitive area would be open to skiers, walkers, and all citizens; but it would be closed to 'extension of road and engineering works, construction of buildings and commercial development'.

That announcement broke with a shattering explosion the Trust's twenty years' silence about the primitive area. Sir William Hudson declared that he was dumbfounded and addressed a hot protest to the Park Trust. Thereupon the Trust reaffirmed its decision and added for good measure that the decision could not be rescinded except by joint action of the three governments which shared responsibility for the Snowy project. But the Commonwealth government would find itself in serious trouble if it accepted its share of that responsibility; for one of its members, Spooner, was an ardent supporter of the Authority, while another, Barwick, was a staunch champion of the Trust. Divisions within the New South Wales government, if they existed, were better concealed: the only public statement on the conflict came from the Minister of Lands, K. C. Compton, who announced that the Authority had no power to proceed with the projected works unless and until it was granted a licence to do so. On that, Sir William Hudson may well have concluded that the political cards were stacked against him. Towards the middle of April he invited the Trustees to meet representatives of the Authority in the disputed area; they accepted, on condition that they could bring with them three members of the Academy of Science. This summit meeting, as the newspapers called it, took place on 19 April and collapsed in angry disagreement. Nevertheless, tempers began to cool within a week or two. Sir William Hudson had already made the decision to tunnel under Townsend Spur into Lady Northcote Canyon rather then deface its walls with a covered pipeline. The Minister of National Development announced on 2 May that there had never been any intention of making an immediate start on the Kosciusko dam. On 7 May the New South Wales Cabinet gave credit to the Park Trust for having done its duty, but considered that more facts were needed before any final decision could be made upon the matters in dispute. Quite a lot of people, it would appear, were seeing virtue in the Byles' doctrine of saving other people's faces.

From this dramatic battle between opposite forces and values the Park emerged victorious. Hitherto, its impact on public opinion had been negligible; but from 1963 onwards its worst enemies had to reckon with it as a power in the land. Its climb to efficiency in the performance of its duty may be illustrated as follows:

Staff employed in the Park

1959	1963	1968
5	26	74

The two landscapes

In 1959 there had been not a single ranger in the service of the Park; in 1963 there were three rangers; in 1968 there was a chief ranger, assisted by seven rangers-in-charge and eleven rangers. The time was now long past when the Superintendent had had to waste his time clearing snow from the road.

The big leap forward was coincidental in time – but not otherwise coincidental – with the advent in 1965 of T. L. Lewis as Minister of Lands. In the war, Lewis had been a commando; in his department he was a rushing mighty wind. He gave orders that the intolerable arrears of departmental business must be cleared up within a year – an impossible task, everybody said; but Lewis achieved it. In the same spirit of assault he shifted the administrative headquarters of the Trust from Sydney to the Park and took the chair himself at the Trust's meetings. At the same time he hammered out a National Parks and Wildlife Policy for the entire State of New South Wales. He knew what he wanted.

> I'm a rationalist about this. Man is an animal, though we tend to forget it. He needs to survive and propagate and have a habitat like any other animal ... I think man would be always clever enough to survive, but he should not take the risk of destroying too much of his environment. You can never put back what you have destroyed.[1]

Those ideas took shape in the National Parks and Wildlife Act 1967. Under this Act the Kosciusko State Park was re-named the Kosciusko National Park. The Park Superintendent was made responsible to the Director of the National Parks and Wildlife Service of New South Wales, who in his turn was directly responsible to Lewis as Minister. The Kosciusko State Park Trust ceased to exist. This did not mean that the Minister had no use for its members; on the contrary, he kept them at work as an advisory committee and regularly took the chair at their meetings in the Park. For the Superintendent of the Park, life became henceforward almost too exciting: in the old days he had kicked against the time-wasting pricks of bureaucracy, but his complaint nowadays – if he had a complaint – was that the Minister hustled him too much.

Lewis, nevertheless, knew how to play a patient game when circumstances required it. In the perennial question of mountain grazing, the need for patience was beyond question. In the early 1960s, when Gare was fuming at the general disregard of the ban on grazing above 4,500 ft, Byles had told him to keep cool – 'This grazing business is all a game of wits and tactics ... Three years ago I forecast that it would be a seven years war. Four years to go.' But there was a longer time than that to go. In March 1965 a disastrous bushfire swept through the high country, including some areas of the Park. Thereupon an agitation was let loose to

---

[1] 'Tom Lewis, M.L.A. – a profile by Evan Williams', *S.M.H.*, 22 August 1969.

cut the Park down to a more manageable size and to encourage 'controlled grazing' as the most effective means of reducing inflammable litter.[1] A succession of drought years followed the bushfire. The Lands Department gave a short reprieve to some graziers whose tenures were running out; other graziers sent their sheep in scores of thousands to graze illegally in the mountains. The rights and wrongs of mountain grazing, both legal and illegal, were debated day by day in letters to the newspapers: 'If a law tells me,' one correspondent wrote, 'that I must watch my animals starve while within walking distance of good fodder, then I'll cheerfully break that law.' Another correspondent answered, 'If graziers knew that they could use the Park in time of trouble, they would naturally bank on it and postpone uncomfortable adjustments.' From this protracted and passionate debate, some onlookers at least drew an uncomfortable conclusion: the times of maximum hardship for the graziers were also the times of maximum danger to the catchments.[2]

From its base in Snowy River Shire, the Snow Lessees Association conducted a spirited campaign. It protested that it wanted *controlled* grazing only. On 21 February 1968 the Premier of New South Wales placated it with a declaration of his government's intention to initiate a thorough and impartial investigation of the whole question of controlled grazing and longer-term leases within certain areas of the Park. On 27 June the Minister of Lands told a deputation that the former Director-General of Agriculture, Dr Graham Edgar, would conduct the investigation. The Edgar Report was submitted in May 1969. It recommended the total abolition of grazing in the Park. The Snow Lessees Association protested that Dr Edgar had been neither thorough nor impartial and called for a new investigation by a five-man committee. All in vain – on 29 May 1969 the Minister of Lands announced the government's acceptance of every recommendation in the report. That announcement put a full stop to 135 years of recorded history.[3]

One has a fellow feeling for the last-ditch defenders of a lost cause. Some of them, no doubt, possessed sufficient capital, skill and determination to make a new start with new methods; but others did not. An old man named Charlie Collins eked out the last years of his life in a hut on the eastern bank of the Thredbo River. In the years of his strength he had worked for the Spencers of Waste Point; in his later years he had earned

---

[1] Reference was made on p. 26, above, to another method of reducing litter, i.e. controlled burning. This subject is too complicated for discussion here, but has been admirably discussed in the Australian Conservation Foundation's Viewpoint Series No. 5, *Bushfire Control and Conservation*. The conclusion is that control burning is necessary in many areas, but its application requires more knowledge, discrimination and restraint than are usually in evidence at present.

[2] See e.g. *Canberra Times*, 12 and 18 February, 6 and 7 March 1969.

[3] If this precision of dating seems pedantic, see p. 132, above.

his living by grazing cattle and conducting riding parties through the high country. Unfortunately for him, nearly all his land lay across the river, within the external boundary of the Park. It was land most beautifully situated; the Park was bound to resume it. Charlie Collins was offered a price that was fair and even generous on a pastoral valuation; but before the land was resumed, a businessman, alert to the prospects of land development, had offered him a higher price. Charlie refused to take less. In the end he got nothing. His death, when it came, was symbolic – but of what? Was it the death of a silly old man, or the death of a folk hero?

Today, Banjo Paterson's man on horseback is a stranger to the Snowy Mountains; but Barcroft Boake's man on skis is making himself at home in them. We may identify three types of skier. First, there is the intrepid tourer, twin brother to the tiger bushwalker; he loves the *langlauf* and the swift glide downhill over powder snow unmarked by anybody else's tracks; scorning comfort, he skis from hut to hut with a heavy pack on his back. Today, he is one skier in ten thousand. Secondly, there is the dashing yet elegant practitioner of swoop and turn on the *piste*; he scorns neither the chairlift nor the hot bath nor the bar nor the good dinner, yet still from time to time finds joy in making the first tracks down the mountainside. Today, he is one skier in a thousand. Thirdly, there is the lover of fun and games with the crowd; he may be a tyro practising the snowplow, he may be a racer practising the slalom, but either way he will never leave the *piste*. Today, he is almost every skier. He is coming to the snowfields in his thousands; before long he will be coming to them in his tens of thousands. Winter sports, which used once to be the privilege of the few, have become in our time accessible to the many. They have become mass recreation and big business. The entrepreneur who wanted to build a chairlift from the floor of the Geehi to the summit of Townsend was not merely a dreamer; no more had been the editor of the first volume of the *Ski Year Book*, as he looked forward to the day when an hotel would nestle at the foot of every snowy peak.[1] Conceivably, that day might come. If and when it did, the proprietors of chairlift and hotel would feel the need for a return on their investment not only in the winter but also in the summer season. Tarred roads would run through the wilderness. Mt Kosciusko would be levelled to make parking space for cars. In New South Wales, as already in some States of America, there would be smog over the National Park.

Do the people of New South Wales want that? Parliament has steadily maintained that they do not want it. Parliamentary legislation, nevertheless, has left some room for manoeuvre. The Park Acts from 1944 on-

[1] See pp. 139, 175, above.

wards have recognised the diversity of human desires and needs. How to satisfy these diverse desires and needs has been the main problem of Park management. In the late 1950s, Byles and Gare started work on a master plan which would identify the areas or zones best suited to this or the other use. In the middle 1960s, a strong committee under the chairmanship of M. F. Day pushed that work towards completion. The chairman and most of his colleagues were scientists; but they sought the aid of laymen – not least the Canberra bushwalkers, who studied and measured the park mile by mile. If ever additional drive was needed to get the work completed, the Minister, T. L. Lewis, gave it. On 29 October 1969, the committee submitted its *Report on the Proposed Park Zoning Plan.* The report's classifications were as follows:

Areas	Paramount management aim
Class I: Wilderness areas. ('wilderness' is the equivalent of the formerly used word 'primitive')	To preserve the natural environment, free from man-made developments.
Class II: Outstanding natural areas.	To preserve outstanding scenic and/or natural scientific values.
Class III: Natural recreation areas.	To provide for appropriate outdoor recreation in natural surroundings.
Class IV: Development areas.	To provide the fullest facilities for recreation and sports, including snowsports, consistent with preserving the Park atmosphere.
Class V: Historic sites.	To preserve, and encourage visits to, points of historic interest.
Class VI: Hydro-electric areas.	Smooth operation of the system and the safety of visitors.

If this plan wins and holds the approval of the Australian people, the story of Monaro will remain for centuries to come the story of two landscapes: one of them predominantly man-made, the other predominantly native.

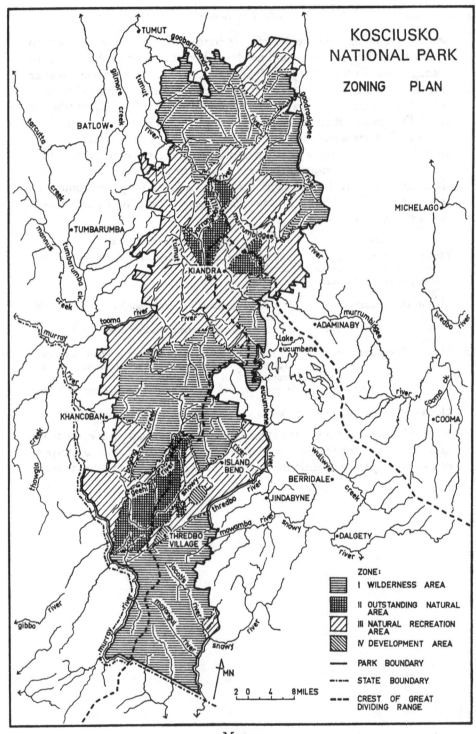

KOSCIUSKO
NATIONAL PARK

ZONING   PLAN

TUMUT

BATLOW

MICHELAGO

TUMBARUMBA

KIANDRA

ADAMINABY

Lake
eucumbene

KHANCOBAN

COOMA

ISLAND
BEND

BERRIDALE

JINDABYNE

DALGETY

THREDBO
VILLAGE

MN

2 0 4   8 MILES

ZONE:

I WILDERNESS AREA

II OUTSTANDING NATURAL
AREA

III NATURAL RECREATION
AREA

IV DEVELOPMENT AREA

——— PARK BOUNDARY

—·—·— STATE BOUNDARY

— — — CREST OF GREAT
DIVIDING RANGE

Map 15

3. Looking forward

Historians, most people say, are students of human experience in past time; but some historians find that statement too simple. A philosophical historian of great distinction, the Frenchman de Jouvenel, refuses to divide human experience into three separate segments of time – past, present and future. Recently, de Jouvenel has established an organisation for the study of experience in the future, in so far as its possibilities are foreshadowed by past and present experience.[1] These possibilities, it need hardly be said, are not easily identified and analysed; yet they are the everyday concern of all of us. As we try to take stock of our individual situations and prospects we say to ourselves something like this: 'Here we are now. How did we get here? Where do we go from here?'

THE TABLELAND

From the mid-nineteenth century up to the present, people in Monaro have earned the larger part of their incomes by producing wool. Their speciality, by and large, has been wool of medium fibre; but for present purposes the graph of average prices (fig. 3) sufficiently illustrates the recent declining trend of their monetary receipts.

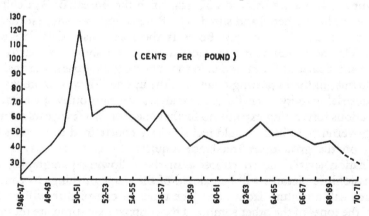

Fig. 3   Australian average greasy wool auction prices, 1946–71

[1] See Bertrand de Jouvenel, *The Art of Conjecture* (Eng. trans. London, 1967).

Fig. 3 tells wool producers a good deal of what they need to know about their present situation. Fig. 4 goes some way towards telling them how that situation has arisen.

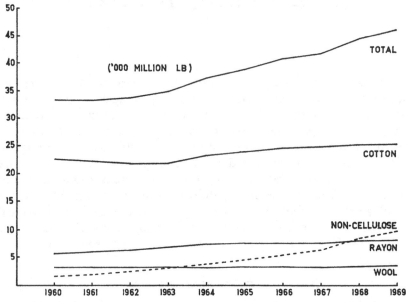

Fig. 4 Estimated world consumption of apparel fibres, 1960–9

Taken together, the two figures make it clear that wool producers, in Monaro as elsewhere, have good reason to feel anxious about their future prospects.

Figures 3 and 4 are based on figures which the Bureau of Agricultural Economics has gathered and sifted. The Bureau does not suggest that the dramatic advance of synthetic fibres is the sole cause of falling wool prices. On the contrary, it pays close attention to short-term factors of supply and demand: for example, in the exporting countries, seasonal ups and downs; in the importing countries, the ups and downs of stocks and of industrial activity. The Bureau's analyses are rigorously quantitative. Its cautious forecasting extends no further ahead than a few years at most. Any governmental body would feel similarly constrained to keep its feet firmly on the ground. A free-lance inquirer, by contrast, may let his imagination soar if he so chooses. In the following paragraphs, two impressionistic pictures will be painted of the shape of things to come in Monaro half a century from now. The tones of one picture will be sanguine, the tones of the other sombre. That contrast may stimulate thought.

The first picture could be entitled 'Business as usual'. It shows Monaro's graziers half a century from now doing very well for themselves in the good old wool trade. On the side of demand, thanks largely to a large investment in research, development and sales promotion, a sufficient number of consumers remain convinced that wool, whether unadulterated or mixed with synthetic fibres, still makes the best dresses and suits. On the side of supply, the costs of classing, packaging, transporting and selling have been cut to an extent which assures for wool a continuing share – not large, but secure – in the expanding world market for apparel. Nevertheless, the graziers in A.D. 2021 are earning half their incomes, if not more, by producing and selling meat. That important development could be traced back to the late 1960s, when the prospects for beef and lamb were looking as bright as the prospects for wool were looking bleak. Thereafter, wool staged a recovery; but 'diversification' still remained the policy of pastoral producers. In consequence, the economy of Monaro now stands on two sturdy legs. The pasture is in good heart. The people are prosperous.

The second picture could be entitled 'The fall of pastoral Monaro'. As could have been predicted from the graphs of the 1960s, synthetic fibres have won their battle with natural fibres; nobody any longer wears wool 'next the skin' or on any other part of his person. What is almost as bad, very few people nowadays eat meat. A cloud which fifty years before had been no larger than a man's hand now darkens the whole sky; 'simulated meats', concocted from soya beans or petrol or heaven knows what, are giving the world's teeming millions all the protein they need at prices they can afford to pay.[1] With both its markets dead or dying, the pastoral economy is derelict. How then is man in Monaro earning his living? Monaro has become an enclave within a monster conurbation. The chain of cities, satellite towns and suburbs stretches along the coast from north of Newcastle to west of Geelong. Canberra by now is more than half-way to Cooma. Across the high mountains, 'decentralisation' – no longer merely a posture – has brought to birth a rapidly growing city. Twenty million or more dwellers in south-eastern Australia are raising their spending power at the rate of 2 or 3 per cent per annum. The tourist trades are booming. They are bringing more money into Monaro than wool and meat ever brought. Hazeldean and Bukalong have become multi-storey hotels; Mount Cooper is a thriving tourist village. Pioneers of the new way of life are providing all the pleasures which city dwellers seek in the vast open spaces – golf courses, heated swimming pools, cocktail bars, poker machines, bingo. From the hotels and tourist villages helicopters take off every morning for the high country: in winter, for the chairlift and the *piste*: in summer, for the flower carpeted alpine meadows – what is left of them.

[1] See the very informative paper by J. L. Sault and J. B. Gale, 'A review of developments in simulated meats' in *Quart. Rev. Agric. Econ.* XXXII (1970), no. 4.

Each of these pictures makes sense up to a point; but each is distorted by the will to believe – or disbelieve. The painter of the first picture has turned a blind eye upon synthetic fibres, simulated meats, crowding tourists and all other disturbers of the 'old days, old ways'. The painter of the second picture has had an eye for the new ways only; but his pretended realism is nothing more than romanticism turned upside down.

Any serious student of the emergent future must pay close attention not only to quantity but also to quality and to the interplay between the two. Consider, for example, what people eat. Nebuchadnezzar ate grass, but

> . . . murmured, as he took the unwonted food,
> 'It may be wholesome, but it isn't good.'

In 1943, when the British meat ration fell to ten pennyworth a week, an elderly Oxford don fell into a rage when somebody told him that soya bean sausages, steamed cod and boiled potatoes were giving him all the calories he needed: 'I don't want calories,' he growled. 'I want chops.' Fifty years from now, simulated meats far tastier than those at present on the market may well be on sale at half the price of the genuine article; but even so, millions of well-to-do Americans, Europeans, Japanese, Chinese and Australians may be clamouring for chops, steaks and hamburgers. Should this prove to be true, the improved pastures of Monaro will still be producing lamb and beef. Perhaps they will still be producing wool.

The upshot of the discussion so far is that the market prospects in long term for pastoral products can as yet only be guessed at. By contrast, the continued massive growth of urban populations within easy reach of Monaro can be predicted, in general if not in detail. Concurrently with this growth, there will be a massive increase of demand to use Monaro for the benefit of people from the cities. Some of these people will buy land, live on it and perhaps improve it; but the large majority will be transient visitors in search of pleasure. Some of the visitors will find what they are looking for on the tableland; they will go fishing, they may go riding – will horse-breeding revive in Monaro? Most of the visitors, however, will seek their pleasure in the high country. Their ideas of pleasure will be various: not all of them will want to see hotels and chairlifts 'at the foot of every snowy peak': some of them will want to preserve a habitat for Burramys, the pigmy possum which until quite recently was known only as a fossil. Conflicts will thus arise among the pleasure seekers. Businessmen, soil conservationists and scientists will become parties to these conflicts. Their outcome will be decided, not merely by the thrust and counter-thrust of sectional interests, but by governmental policy, arrived at in public debate.

Looking forward

THE HIGH COUNTRY

In what follows, policy in the Park will be the focal point of discussion. The principles of policy have been declared in two important Acts, the Kosciusko State Park Act 1944 and the National Parks and Wildlife Act 1967. Under the first Act, implementation of policy was the responsibility of trustees and of their chairman, the Minister of Lands; under the second Act, it becomes the responsibility of the National Parks and Wildlife Service, working under the direction of the Minister of Lands.

Today, the tasks which challenge policy makers and administrators are immensely different from what they were a quarter of a century ago. Then, the grazing interest was still predominant in the high country; today, it has no place at all within the boundaries of the Park. Then, the Snowy Mountains Authority was just beginning to plan its mighty works; today, those works are completed. Then, snow sports were still a recreation of the few; today, they are big business for a mass market.

No economic check can be foreseen to the growth of demand for mass recreation in the Park and to a corresponding growth of the supplying services. The check is political. Had it not been operating, the tourist industry would already be more widely scattered than it now is:[1] were it to be removed, the industry would sooner or later make a strong bid for all the snowfields within the Park's boundaries. The zoning plan of October 1969, with its strict demarcation of 'development areas'[2] demonstrates the strong determination of some Australians to resist any such bid. But why should any check be imposed upon the freedom of the tourist industry to locate itself as it sees fit? Tourists, together with the large and small businesses which supply their needs, constitute a legitimate interest. Moreover, the idea is still widely prevalent in Australia that every individual, in pursuing his own interest, is led by an invisible hand to promote the interest of his neighbours. If that has been the highway to wealth and welfare elsewhere, why not also in the Park?

Not so many people nowadays would give as confident an answer to that kind of question as their grandfathers gave. Too many instances can be seen of men in lawful pursuit of their own interests doing hurt to the interests of their neighbours. Sometimes, a large industry in pursuit of its own interests does hurt to the entire community. In contemporary economic analysis, a distinction is commonly drawn between private and social benefits, private and social costs. In the formulation of Park policy, this distinction is basic. The men who frame policy are bound to consider

[1] See, e.g., p. 181, above.
[2] The 'development areas' shown on map 15 are to some extent provisional. The *Report* stated that additional areas were being investigated. At the same time, it stressed the need for tourist townships *outside* the boundaries of the Park.

not only the benefits which accrue to a single powerful group of people, but also the costs which fall upon smaller groups of people whose interests are equally legitimate and, possibly, equally important to the community as a whole. To render justice to all these groups is the Park's most difficult task.

When the grazing interest was dominant in the high country, Richard Helms put his finger on this problem of distributive justice.[1] He thought it wrong that one section of the community, in pursuit of its own interest, should be permitted to endanger the community's most important water catchment; he thought it equally wrong that this section should be allowed to despoil 'an unsullied alpine landscape, and to replace a fresh and fragrant growth by dead and half-burned sticks, making a desert of what was once a garden'. He demanded justice for the minority groups of his time, such as tourists, artists, scientists. In our time, tourists have become the majority; but the rights of minorities are the same now as then.

All the minority groups have one thing in common: they consist of people who take their pleasure non-gregariously. Poets who frequent the Park could perhaps be counted on the fingers of one hand; but two of them at least, Douglas Stewart and David Campbell, have enhanced the quality of Australian life. In comparison with the poets, the bushwalkers are quite a large group. They played a decisive part in getting the Park established; it would seem only fair to leave them some bush to walk in. The scientific field workers constitute a group of intermediate size; like the poets and the bushwalkers, they need the zoning plan, or something equivalent to it. Their determination to resist further encroachment upon the wilderness sometimes provokes angry and contemptuous protests – why should progress be held back in order to save the David moraine or to preserve a habitat for Burramys?

Instead of answering this question, some scientists emphasise the utilitarian value of their work in the Park. Protection of the mountain catchments, for example, is both a task of applied science and a vital national interest. To cite another example: preservation of the Park's rich and diverse flora is quite possibly a vital interest not only of the Australian people but of the human species everywhere. A distinguished botanist, Sir Otto Frankel, has written a series of papers to explain how and why this is so. Today, even more than in the past, higher productivity of crops and pastures is sought by specialisation on a few plants. Concomitantly with the specialisation, the risks of disaster through plant disease are increased: what happened last century in Ireland when the potato crop failed could happen again as 'the green revolution' of this century gathers pace. In their endeavour to fend off so great a disaster, plant breeders have to find a counterpoise to specialisation. They find it in the rich variety of the gene

[1] See p. 144, above.

pools of original nature. Seen in the light of Frankel's exposition, Monaro's two landscapes acquire an enhanced significance. On the tableland, the original native variety of the flora is being eliminated. In the high country, a precious reservoir of genetic diversity can still be preserved.

The utilitarian argument, as exemplified above, appears cogent. There emerges from it the following unchallengeable conclusion. Any realistic appraisal of competing interests requires a good deal more than a mere counting of hands in each separate group of Park users: if heads must be counted, the heads of farmers in the irrigation settlements downstream and of water consumers in far-distant Adelaide will need to be included. So far so good; but the utilitarian argument will collapse it if is pushed too far. The David moraine and Burramys will not be saved by attempting to prove them useful. If they do survive, it will be because Australians reject utility as the sole criterion of value.

Utility and interest should not be, and in fact are not, the sole determinants of policy. Policymakers are not calculating machines but human persons, with their own personal preferences and values. Explicitly or implicitly, every debate on policy is a debate about values. In a democracy, the debate is continuous. It involves everybody. Even the man who says 'I don't care' is party to the debate; he is giving his vote for a policy of drift. The man who says 'Leave it to the future' is doing the same. The future is now. What we do now or fail to do now is making the future or wrecking it. We are deciding here and now whether or not Australian fauna and flora of the high country are to possess a habitat in which they can survive. Many Australians contend that this habitat must at all costs be safeguarded.

Such a contention is a far cry from the doctrine that man is 'monarch of all he surveys . . . lord of the fowl and the brute'. In the long history of theological and philosophical reflection on man's place in nature, this has been the majority doctrine, at any rate in the West. Until quite recently, very few people have contended that man has a moral obligation to respect and preserve nature's original diversity. On the contrary, influential philosophers, most notably Descartes, have persuaded themselves and their disciples that nature is merely a mechanism to be manipulated by man. To be sure, that doctrine has proved vulnerable to evolutionary theory. Western Man is learning at long last that he is a member of the natural world, not merely its master. Increasingly, he is becoming aware that his powers as master, if he exercises them imprudently, may endanger all life on this planet, his own life included. Prudence, however, is not the sum total of philosophy. General J. C. Smuts, among others, has explored the moral implications of man's duality. That exploration, however, will need to be carried much further if ever conservationist programmes are to be given a firm anchorage in philosophy.

Meanwhile, individual intuitions and ideas of right and wrong must make good as best they can the present imperfections of philosophical guidance. The forester Byles is as helpful a guide as can be found to the issues which call for decision in the Kosciusko National Park. In an address in April 1964 to forestry students, he drew a contrast between the simple duties of foresters in earlier times and their duties today, when 'multiple use' means, among other things, the production of timber, the protection of wildlife and the provision of facilities for recreation. The life of a fores-ter, he said, would certainly be a good deal simpler if forests were enclosed within man-proof fences; people start fires that have to be put out; they scatter litter that has to be cleaned up. People, however, have come to stay. 'So you will just have to learn to live with your public,' Byles told the students, 'look after it, educate it, help it to enjoy itself and expect to get your *quid pro quo* in the form of support (from a section of it at any rate) for your forestry programmes.' The same tolerance finds expression in an address which Byles gave a year later to the Canberra Alpine Club. The National Park, he told his audience, contained:

> everything that one could wish for in the way of natural features: high alti-tudes and low altitudes: cold climates and hot climates: steep rocky cliffs and gentle rolling plains: dense forests and open meadows: trees that grow 180 feet high and miniature shrubs that dare not raise their heads more than three inches above ground level: rushing torrents and meandering brooks: caves in which one cannot see one's hand in front of one's face and hundred-mile vistas of range after range into infinity. It contains birds, animals and reptiles of every size and type . . .
>
> These assets are there to be enjoyed but not destroyed: to be used but not abused: to serve the present generation and to be preserved for future genera-tions.

But how, he asked, could this be achieved, seeing that human beings had such diverse preferences and values? Some loved crowds, others loved solitude; some enjoyed roughing it, others enjoyed comfort; some cared for the birds, animals, insects and plants, others did not care. All these people, he insisted, possessed the right to enjoy themselves in the ways they preferred; but no section of the people possessed a pre-emptive right over the entire Park.

Everything that Byles has written about the Park reveals his individual preference for strenuous walking, assiduous research and solitary con-templation. His contemplative faculty finds characteristic expression in a paper entitled *Snow Gum – the Tree*. This tree, he says, has never earned dividends for timber companies; but it conserves soil and water. He explains in precise detail how it performs this function. So far, his explana-tions have been scientific; but a change takes place in their character and style.

We cannot appreciate anything fully until we understand it, until we pick up its wavelength so to speak, until we learn to think the way it thinks . . .

So, if we wish to appreciate this particular Australian tree we must try to understand its point of view, realising that it is a living organism like you and me . . . We must try to understand its manner of living, its philosophy of life, its place in the world of natural things and the spirit that keeps it going in spite of great adversity.

Those sentences call to mind Douglas Stewart's poem, 'The Snow-gum'. They bridge the gap between scientific reasoning and aesthetic sensibility.

Two short sentences complete the statement of philosophical or religious belief which is present by implication in everything that this old forester has written or said about the high country. These sentences have a Homeric ring. Byles wrote them at the climax of the battle between the Park and the Snowy Mountains Authority.

All the water in the high levels of the Park must not be converted to power. *Some* of it must be left on the altar of the gods.

This credo has taken political shape in the achievement of T. L. Lewis as Minister of Lands and as creator of the National Parks and Wildlife Service of New South Wales. Throughout the past eighteen months, the Minister and the Service have been at work upon a plan of management for the Kosciusko National Park. If no unexpected hitch occurs, the plan will be in operation before this book appears in print. It will render justice to all legitimate interests in the Park – not least, to the skiers and the summer tourists. Nevertheless, the expansion of the tourist industry will take place predominantly *outside* the Park's boundaries. Those boundaries, it seems safe to predict, will be extended as opportunity offers. Within them, zones will be demarcated. Spacious wilderness areas will safeguard for as far ahead as can be foreseen the high country's inheritance from primaeval time.

APPENDIXES

Bukalong rainfall records

Note: For the records 1858–1944 see *Results of Rainfall Observations NSW* (Commonwealth Bureau of Meteorology, 1948), p. 167. For the records 1945–70 I am indebted to Mr C. T. Garnock. In addition, Mr Garnock has also given me temperature records, which suggest some interesting correlations with the rainfall.

Year	Jan	Feb	Mar	Apr	May	Jun	Jul	Aug	Sep	Oct	Nov	Dec	Total
1858	0.38	3.63	1.63	1.13	3.00	1.25	2.00	1.75	2.75	2.75	5.50	3.50	29.27
1859	3.50	0.88	0.50	0.50	0.75	2.38	0.25	0.63	4.13	0.63	2.50	4.00	20.65
1860	4.00	5.75	2.25	1.75	0.25	1.75	9.50	0.75	1.00	2.25	6.00	2.50	37.75
1861	2.75	5.00	0.13	1.38	1.50	0.25	1.38	1.50	0.63	1.13	2.63	1.25	19.53
1862	1.00	2.00	0.88	3.50	1.63	1.25	0.63	1.75	0.63	0.88	0.75	1.75	16.65
1863	2.75	3.75	2.25	0.75	0.25	4.50	0.63	0.75	1.75	2.50	3.25	2.75	25.88
1864	0.88	1.25	3.88	2.88	0.50	3.25	4.50	1.88	2.13	3.00	0.50	3.00	27.65
1865	2.50	0.25	0.75	0.25	1.63	0.25	2.00	1.50	0.63	0.63	1.00	0.88	12.27
1866	4.00	1.38	1.00	0.88	2.63	2.13	1.13	0.75	1.25	2.00	2.38	5.50	25.03
1867	2.38	2.75	4.13	5.50	2.63	6.00	0.88	0.75	2.13	1.88	0.38	0.63	37.29
1868	3.00	3.25	0.25	0.13	9.88	0.50	1.75	0.50	0.25	1.38	3.13	1.00	15.89
1869	3.38	3.63	1.25	4.00	0.75	1.63	0.13	0.63	2.25	3.50	2.00	2.13	29.66
1870	2.88	0.25	10.13	8.38	7.88	1.75	2.75	1.13	1.75	5.00	4.75	4.13	50.78
1871	4.00	4.25	2.88	5.38	10.13	7.75	0.25	0.25	3.50	3.75	2.50	3.75	48.39
1872	2.63	1.63	2.13	1.75	0.88	1.00	1.63	1.13	2.63	1.88	4.50	2.50	24.29
1873	6.38	12.63	1.00	6.13	0.75	6.75	1.75	1.00	1.88	1.75	7.50	0.75	48.27
1874	3.38	7.25	5.38	0.75	2.25	5.63	4.00	1.13	2.13	1.50	2.13	1.50	37.03
1875	1.00	3.13	2.38	3.13	2.38	3.63	3.25	1.75	2.13	2.50	1.50	1.63	27.66
1876	2.50	1.38	1.38	0.63	1.88	0.75	0.88	0.75	2.38	6.00	4.38	0.50	25.78
1877	0.25	1.13	0.88	1.63	2.50	1.13	0.88	0.63	3.88	2.88	0.50	1.88	18.17
1878	0.38	7.13	2.00	0.38	0.13	1.38	0.88	0.63	2.13	1.75	1.50	1.75	20.04
1879	2.58	3.03	2.14	2.33	7.50	0.75	1.50	2.75	5.25	3.75	2.25	1.63	35.46
1880	1.25	4.63	4.25	4.38	3.38	1.63	3.38	1.25	1.88	1.88	2.50	1.99	32.40
1881	0.99	3.06	3.32	1.06	1.60	3.02	0.20	0.60	1.93	3.93	4.75	2.43	26.89
1882	1.03	0.60	1.82	1.78	1.45	1.75	1.48	1.08	0.72	4.04	2.21	1.74	19.70
1883	2.41	2.94	0.98	2.61	1.75	0.54	0.77	0.39	1.57	2.34	2.65	0.77	19.72
1884	1.76	0.88	1.61	5.93	0.47	1.06	0.60	1.18	0.87	3.13	1.33	0.66	19.48
1885	1.88	1.65	1.04	0.76	0.14	0.66	0.57	0.27	1.36	0.71	3.42	0.77	13.23
1886	4.40	2.42	0.29	1.59	0.37	0.12	0.47	1.62	1.78	2.70	1.68	3.85	21.29
1887	4.46	2.08	2.61	1.15	1.20	2.70	1.52	1.46	2.63	2.48	4.21	4.60	31.00
1888	2.05	2.31	1.05	0.00	1.11	0.17	0.35	3.32	1.90	0.57	1.15	7.70	21.68
1889	5.27	2.53	0.05	0.78	3.38	2.63	0.63	0.59	1.30	2.85	3.93	0.71	24.65

Year												
1890	1.13	4.03	3.30	1.29	3.16	6.25	2.37	1.13	1.43	2.31	1.74	2.02	30.16
1891	3.63	1.28	0.74	2.05	0.47	15.79	6.17	4.15	3.23	2.30	3.38	2.19	45.38
1892	1.96	0.74	3.11	3.42	0.94	0.30	1.27	1.61	5.88	6.19	3.33	1.43	30.18
1893	4.14	0.67	2.94	4.27	1.33	1.85	3.04	0.23	2.96	3.13	4.03	7.34	35.93
1894	1.40	2.44	4.43	2.20	0.39	4.37	0.13	1.25	1.52	1.64	0.14	2.23	22.14
1895	2.31	1.83	2.49	0.74	0.57	2.31	0.84	1.53	1.21	0.18	0.00	1.41	15.42
1896	1.77	2.51	0.79	0.44	3.73	6.55	0.34	2.19	2.52	0.69	3.81	2.21	27.55
1897	4.00	75.8	1.36	1.53	2.10	1.68	0.83	1.63	2.18	1.28	0.51	2.40	27.31
1898	0.07	81.8	1.04	0.25	1.42	1.41	1.64	2.19	0.80	3.76	0.36	1.55	20.37
1899	2.27	0.48	1.74	6.61	1.23	3.15	2.81	3.22	0.93	1.26	1.40	0.26	25.36
1900	1.85	0.45	2.76	4.41	6.87	1.46	2.91	1.04	2.04	0.35	1.60	1.29	27.03
1901	1.70	0.38	1.36	2.35	0.06	1.79	2.01	4.79	0.92	2.20	1.88	0.25	19.69
1902	3.21	0.26	2.73	0.87	0.24	2.03	5.61	0.37	1.61	1.86	0.97	4.89	24.65
1903	0.11	0.99	1.57	0.90	1.53	1.58	1.62	0.54	2.72	2.84	1.07	2.29	17.76
1904	3.86	2.39	0.47	0.92	1.05	0.68	1.47	2.01	1.15	1.23	0.81	1.14	17.18
1905	2.36	1.44	3.80	1.82	1.22	1.25	1.08	2.32	0.20	5.37	0.07	1.43	22.36
1906	0.06	1.05	6.60	0.23	0.35	0.44	0.34	1.83	1.59	2.24	3.11	1.84	19.68
1907	3.94	1.07	1.56	1.38	0.87	2.78	0.00	0.68	0.28	0.58	2.10	4.12	19.36
1908	3.20	1.33	0.64	1.46	0.41	1.03	0.29	3.66	3.46	1.10	3.52	1.52	21.62
1909	3.01	2.94	0.72	0.13	0.16	4.19	2.51	0.50	0.77	1.32	0.25	1.60	18.10
1910	5.72	0.00	1.80	0.13	1.42	0.29	1.58	0.07	2.50	1.46	1.76	2.47	19.20
1911	7.85	2.86	4.33	0.44	1.42	2.20	1.75	1.00	2.03	1.72	0.61	3.10	29.31
1912	0.32	0.80	2.14	0.66	0.47	0.96	4.26	0.25	0.43	0.62	2.07	3.52	16.50
1913	0.74	0.79	5.54	0.95	6.86	6.79	1.08	0.82	1.47	2.39	2.06	0.96	30.45
1914	1.27	0.16	5.82	1.35	0.90	0.63	3.64	0.27	1.80	0.59	1.35	3.73	21.51
1915	2.28	0.00	1.03	0.56	0.57	1.72	0.53	1.30	3.06	1.69	0.60	2.92	16.26
1916	2.50	3.12	1.85	0.94	0.48	1.00	1.38	0.92	3.13	2.44	2.95	3.04	23.75
1917	4.75	1.28	0.55	1.03	0.73	0.87	0.43	0.15	1.67	1.74	2.52	2.25	17.97
1918	3.78	2.32	2.35	1.34	0.33	0.91	2.76	1.10	1.21	1.27	0.96	0.69	19.02
1919	0.00	9.94	1.11	2.10	2.66	0.84	0.15	3.52	0.82	1.26	1.51	5.47	29.39
1920	7.17	1.50	3.21	1.49	0.29	0.52	1.27	1.41	1.36	3.15	0.91	6.70	28.98
1921	1.70	7.26	3.10	3.29	1.39	0.60	0.55	0.50	0.95	2.65	1.57	4.43	27.99
1922	2.78	3.56	0.82	0.30	0.72	1.48	8.38	2.00	3.23	1.80	0.13	1.77	26.97
1923	0.95	0.00	1.05	0.02	0.42	1.47	1.67	0.50	4.40	1.88	2.11	3.18	17.65
1924	3.66	2.90	1.43	1.61	1.02	1.06	1.81	0.92	1.01	1.09	3.51	6.30	25.32

Appendix 1

Year	Jan	Feb	Mar	Apr	May	Jun	Jul	Aug	Sep	Oct	Nov	Dec	Total
1925	2.18	1.02	1.27	2.50	11.22	1.34	3.44	1.37	0.60	2.39	2.75	0.55	30.63
1926	4.26	0.00	2.04	1.67	1.18	2.26	0.93	0.60	1.71	0.96	1.19	0.91	17.71
1927	3.52	0.35	0.73	1.93	2.87	0.00	2.33	0.56	2.15	4.18	1.49	0.60	20.71
1928	1.37	5.19	7.35	1.79	1.29	4.07	0.43	0.36	1.40	0.84	0.46	1.91	26.46
1929	0.29	7.42	0.83	0.96	1.70	0.88	0.28	4.86	1.02	1.33	5.44	2.74	27.74
1930	0.33	2.39	0.46	0.46	1.35	4.49	0.69	1.00	1.02	2.63	0.96	4.13	19.91
1931	1.15	0.69	2.38	1.41	2.58	2.57	1.14	0.78	1.29	2.37	1.85	0.76	18.97
1932	0.09	2.49	2.95	1.74	2.52	0.16	5.89	2.24	1.83	1.85	2.32	2.32	26.40
1933	2.07	0.28	0.73	0.57	1.30	4.01	1.93	0.83	1.66	2.64	2.63	4.13	22.78
1934	11.00	2.12	2.62	3.63	0.40	3.10	4.16	3.96	1.30	3.08	4.16	1.22	40.75
1935	3.00	6.43	0.65	4.58	0.39	0.50	0.55	0.74	1.41	2.51	1.39	3.53	25.68
1936	2.93	2.43	1.44	2.51	0.26	4.40	1.15	1.12	0.67	1.17	1.58	4.46	24.12
1937	2.30	0.86	2.18	0.50	1.08	2.91	0.64	1.31	1.60	4.00	1.05	3.20	21.63
1938	4.23	1.28	2.17	0.53	0.38	0.87	1.42	3.48	2.02	1.60	1.98	0.16	20.12
1939	1.80	1.78	4.08	3.11	0.55	0.50	0.45	4.49	0.50	1.53	2.57	0.27	21.63
1940	2.45	0.80	0.00	5.16	1.30	0.70	0.00	0.41	2.92	0.16	2.10	3.21	19.21
1941	4.19	0.51	0.88	1.15	0.40	1.50	0.25	0.62	1.20	0.35	1.83	0.87	13.75
1942	0.12	3.00	3.55	0.71	1.28	1.32	1.00	0.10	1.42	2.87	4.25	0.25	19.87
1943	5.05	0.45	0.28	0.91	1.90	1.16	0.36	3.80	1.25	3.61	1.47	2.27	22.51
1944	0.30	0.00	1.58	4.90	7.40	0.29	0.15	0.65	0.22	2.09	0.95	1.41	19.94
Averages for 87 years	2.60	2.46	2.12	1.92	1.93	2.18	1.71	1.39	1.80	2.16	2.19	2.38	24.84
1945	4.24	1.31	1.61	3.83	1.66	2.90	0.59	0.68	0.45	2.85	2.32	1.24	23.68
1946	1.49	3.07	3.44	1.38	2.93	5.93	0.38	1.03	1.04	1.32	2.99	2.56	27.56
1947	0.65	4.32	3.12	2.48	0.12	0.89	0.91	1.24	0.89	1.48	2.40	5.14	23.64
1948	4.67	2.25	1.23	2.96	2.14	0.35	0.42	0.00	2.27	2.80	1.81	4.32	25.22
1949	3.39	1.76	3.70	0.25	3.09	6.37	1.79	0.89	1.66	2.76	5.05	1.49	32.20
1950	2.57	6.97	6.85	4.06	2.58	1.83	1.91	1.31	1.10	6.88	3.62	2.44	42.12

Year													Total
1951	1.29	5.20	0.11	1.66	0.24	4.97	0.98	3.12	4.68	3.16	2.61	1.68	29.70
1952	0.47	1.80	3.35	6.06	1.65	9.07	1.69	2.00	0.61	4.37	6.07	6.08	43.22
1953	1.41	2.23	1.17	0.50	8.10	1.29	0.69	2.82	1.06	1.68	1.67	1.58	24.20
1954	3.45	2.76	0.00	0.84	0.51	3.34	0.84	0.50	0.95	1.97	5.45	0.62	21.23
1955	1.65	3.48	2.64	0.36	2.38	1.80	0.50	0.38	1.08	2.85	0.97	6.53	24.62
1956	7.60	4.72	4.58	3.55	5.16	5.63	2.97	1.21	2.02	2.15	1.09	0.66	41.34
1957	0.23	1.81	1.92	0.30	0.80	2.77	6.18	4.44	0.97	1.54	1.49	1.23	24.98
1958	2.76	3.10	1.00	1.11	1.01	2.63	1.39	0.80	2.66	2.12	1.81	1.74	22.13
1959	1.16	2.16	2.37	1.73	0.39	4.72	3.92	1.43	3.56	7.17	3.64	1.74	33.99
1960	4.42	0.57	2.41	2.19	1.83	1.95	5.10	0.78	3.46	1.92	1.83	3.95	30.41
1961	1.71	1.12	8.46	1.24	1.02	1.75	3.73	1.97	3.92	1.54	3.87	2.59	32.92
1962	5.90	2.12	0.66	1.06	1.85	0.20	0.66	0.70	2.42	1.29	1.18	4.29	22.33
1963	3.85	1.80	1.56	2.26	4.07	2.14	3.46	1.10	1.33	1.46	2.50	3.23	28.76
1964	0.12	1.48	1.14	5.24	1.35	1.56	1.04	3.19	1.13	2.54	1.92	2.35	23.06
1965	0.62	0.99	0.25	1.09	0.44	0.53	0.77	4.29	1.12	1.45	2.37	1.11	15.03
1966	1.95	0.78	3.77	0.51	0.64	2.71	1.66	0.66	1.21	3.29	3.94	4.17	25.29
1967	2.56	0.36	1.27	0.16	1.21	0.95	0.79	3.24	3.05	1.45	1.18	1.26	17.48
1968	1.16	0.00	2.47	1.02	2.83	1.18	1.23	0.81	0.12	0.83	1.41	4.76	17.82
1969	1.38	2.79	2.49	2.67	2.89	3.00	0.73	1.00	0.74	1.90	8.07	1.65	29.31
1970	2.32	4.04	4.68	1.60	2.71	1.75	0.50	1.91	1.59	1.34	4.53	5.45	31.97
Averages for first 100 years 1858–1957	2.59	2.56	2.18	1.96	2.00	2.36	1.69	1.41	1.75	2.23	2.28	2.44	25.45
Averages for 113 years 1858–1970	2.56	2.46	2.13	1.93	1.99	2.31	1.71	1.44	1.79	2.23	2.36	2.50	25.45

Tasks

This book is synoptic. It follows a broad historical highway, well supplied with signposts to tracks which ecological, economic, sociological and philosophical inquirers may find it worth their while to explore. For their benefit, I shall cite a few examples.

Historical-ecological inquiry

Patient and precise exploration of small, well-defined areas is called for. For example:

1. On the tableland, the river basin of the Umaralla and its headwater streams.

One could well start with the field books of the mid-nineteenth century surveyors, who mapped the river chain by chain. One would thus get a good idea of its state when the impact of pastoral settlement was still slight. Thereafter, one could draw maps of land occupation at successive points in time, as has been done in the Bradley series in this book. The question would then arise: 'What impact have the occupiers at successive stages had upon their environment?' To answer this question, other maps – of rock and soil, elevation, rainfall, vegetation – would be required. The historical and ecological studies would need to be closely intermeshed.

2. In the high country, a comparably well defined area is the zone of 'wilderness' (see map 15, p. 182) which stretches north from Gungartan and Schlink's Pass.

Here, the mapping of leasehold tenures at successive points in time has already been carried a good distance by Mr Dan Coward. Moreover, ecologists of S.M.A. and C.S.I.R.O. have done a good deal of work within the area, or reasonably close to it. A useful next step would be to establish and to explain the different intensities of grazing in successive periods of time, and closely to examine the consequences.

Both the above-mentioned research tasks, like many others which could be named, require an historian who is able and willing to work in very close contact with natural scientists on the one side, and with economists on the other.

Historical-economic inquiry

Nowadays, economic historians are tending more and more to employ sophisticated theory and well ordered statistics, where the latter are available. Unfortunately, as was shown in the first chapter of this book, the statistical picture of Monaro is a jumble, at least up to the late 1880s, when Sir Timothy Coghlan began to bring order into the official statistics of New South Wales. We need therefore feel no surprise that the economic history of Monaro, and of comparable Australian regions, has hitherto proved unattractive to research workers. Nevertheless, unofficial statistical evidence to some extent makes good the shortcomings of the official data. Moreover, the problems are important. For example:

1. The history of transport.

It was taken for granted in the 1830s and 1840s (see pp. 7–8, above) that the tableland would be linked to the coast. As late as the 1870s, as is apparent in Edward Pratt's diary, there was level pegging between coastal shipping and land transport. Thereafter, the little ships and the little ports rotted. A counter-factual study, of the kind now fashionable in America, might spotlight the benefits and costs, not only in Monaro but throughout the greater part of New South Wales, of a rail and road system dominated by Sydney. Historical research could elucidate the causes, both economic and political, of this dominance.

2. The economic viability of pastoral properties, as determined or influenced by their size.

This question is currently the close concern of agricultural economists. If they were to project their studies of it into past time, or if historians would work forward from (say) the 1860s to present time, more useful answers to the question would be gained than are at present forthcoming. The viable area has been, and still remains, variable, in accord with the varying circumstances of place and time – not to mention individual capacity and character.

Historical-sociological inquiry

In the text of this book, various paths of inquiry have been signposted. For example:

1. Class structure, and mobility between classes.

In the squatting period and for some time afterwards there was a good deal more mobility than historians have usually recognised. Today, mobility is more severely restricted: for example, it is far more difficult than it used to be for a manager to graduate to ownership. What have been the causes, and what the consequences of this change? Here we have a research task worth tackling.

2. The emotional component of land ownership.

Some owners, both large and small, have begun to feel, not merely that the land belongs to them, but that they belong to it (see pp. 162–3). The emergence of this feeling can be traced in family papers: for example, Edward Pratt's daughter (Mrs E. A. Ferguson of Myalla) gave poetic expression to it. Here we have the theme, not of a thesis – God forbid! – but of an essay, which would be comparable in tone with Dame Mary Gilmore's picture of 'old days old ways' in the Riverina.

Historical-philosophical inquiry

The theme of this book was introduced in chapter 1 as a question: 'How has

Appendix 2

man in Monaro used the land on which he lives?' To look for answers to such a question, in Monaro or in any other region of the world, is not merely a task of fact collecting. The inquirer needs also to have in his head clear ideas about man's place in nature; in so far as he does not possess these clear ideas, his explanations and evaluations of human action will be muddled. Philosophers must play their part in clearing up the muddles. For example:

1. Ecological theory (see pp. 61–2) needs, but does not yet possess, a firm philosophical foundation.

2. National parks and wildlife policies (see pp. 187–91) require, but do not yet possess, the same foundation.

So far, philosophers have not been very helpful in these and related areas of thought. This, perhaps, is one reason why the current debate about environmental quality so often produces more heat than light.

Historical materials

The records of Monaro are rich. Among them are some superb station records – Bibbenluke and Burnima, Bukalong and Dangelong – of which only occasional use has been made in this book: moreover, a great deal can be learned from the records of businesses in the townships, as well as from conversations with their present owners. From the 1850s onwards, the newspapers are a valuable source; in using it, the well-indexed Perkins Papers are a great help. Many other sources have been cited in this book. This rich and varied historical material is not the exclusive property of professional historians. Why should not landowners explore the history of their own properties and families? And why should not school teachers bring the explorer's spirit into their classrooms? For example, boys and girls at school in Berridale could be set the task of identifying the people buried in the old churchyard at Gegedzerick, of tracing their descendents still living in the neighbourhood, and of recording their present economic and social condition. That exercise might awaken a lively and continuing interest in the past, present and future of Monaro.

During the past three or four years, valuable historical materials, hitherto unknown to archivists, have been brought to light and their safekeeping assured. People in Monaro will still find it rewarding to search their own homes and offices for letters, diaries and business records. Their discoveries will enrich Monaro's historical tilth.

In the Cooma district, a search is now under way for the burial place of Biggenhook (see pp. 15, 67–71, 112). Meanwhile, archaeological research into the Aboriginal past gathers momentum. So does the political movement for justice to the Aboriginal people. In Monaro they no longer survive; but in other Australian regions they are resurgent.

Index

NOTE. Cattle, sheep, wool, wool prices and similar categories have been excluded from this index because entries would have been needed for every second page. Particular breeds of plants and animals, both native and immigrant, have also been excluded because of the ramifying details. Some exceptions have been allowed, such as rabbits, llamas and Burramys. The last-named has symbolic value.

Index

Barwick, Sir Garfield, 174, 177
Bauer, Ferdinand, 54
Berridale, 84n., 202
Bibbenluke, 10, 91, 103, 106, 135, 155, 159, 160, 202. *See also* Bradley; Edwards
Biggam, 158
Biggenhook, 15, 112, 200
Billylingara, 109
Black Lake, 160
Blyton, J., 101
Boake, Barcroft, 137, 138, 167n., 180
Boers, 44
Bogong moths, 10, 17, 21, 22, 131
Boloco, 109
Bombala, 81, 91, 112, 124
botanists, *see* Banks, Bauer, Brown; Cambage; Cunningham; Froggatt; Howitt; Jackson; Maiden; Moore; von Mueller; Patton
Boucher, John, 81, 159. *See also* Bukalong rainfall records
Bourke, Governor Sir Richard, 7
Boyd, Benjamin, 7, 46, 48
Bradley, William, 41, 44, 48, 49, 50, 78–81, 91, 92, 102, 123, 128
  succession to, pt. III 1 *passim* 155, 116, maps 9, 10, 11, 12, 14
Braidwood, 39, 42, 75
Brassies range, 135, 142
Bredbo, 109
Brindabella mountains, 21, 65, 133
Brisbane, Governor Sir Thomas, 8n.
Brodribb, W. A., 39–41, 48, 78–80, 133, 134
Brookes, Richard, 46
Brown, Robert, 54, 61
Browne, W. E., 164–9
Bruce, Alexander, 109, 122, 123
Bureau of Agricultural Economics, 153, 184
Bukalong, 81, 159, 162, 185, 202. *See also* rainfall records
Burnet, Sir Macfarlane, 14
Burnima, 38, 103, 106, 156, 160, 202
burning
  by Aborigines, 20–6, 63, 131, 132
  by white people, 26–8, 63, 132, 146, 147
Burramys, a pigmy possum, 186, 188, 189
Burrill Lake, 18
bushfires and bushfire prevention, 26n., 63, 166, 174, 176, 178, 179n.
bushrangers, 60
bushwalkers, 166, 188
Byles, B. U., 132n., 144–47, 168, 172–5, 177–80. *See also* alpine catchments; forests; 'hands and knees observation'; Park; zoning plan

C.S.I.R.O. (Commonwealth Scientific and Industrial Research Organisation), 120, 157, 158, 160, 200
Cambage, R. H., 24, 27
Campbell, David, 188
Campbell, Robert, 158
Campbell, Sarah, 159
Campbell, Sophia, 159
Campbelltown, 12
Canberra, 8, 21, 82, 109, 185
Canberra Alpine Club, 181, 190
Carbon Johnny, an Aborigine, 70
Carlaminda, 109
Catchment Areas Protection Board, 171. *See also* alpine catchments; Soil Conservation Service
census returns,
  of Aborigines, 47, 112
  of Monaro, 41, 85, 95
Charley Tara, an Aborigine, 70
Chesterman, A. H., 141
Chisholm, Caroline, 48
churches, 84, 125
Clarke, Rev. W. B., 81, 143
Clayton, E. S., 168, 169, 170, 171, 172. *See also* Soil Conservation Service
Closer Settlement Acts, 106, 155. *See also* Soldier Settlement Acts
Cobb, Nathen, 128
Coffey, John, 99
Colbert's Forest Ordinance, 58
Collins, Charlie, 179, 180
company ownership, 127
Compton, Hon. H. C., 177
Conservation, Ministry of, 171n.
conservationist philosophy, 187–91, 202. *See also* ecology
conservationist practice, pt. IV *passim*. *See also* Byles, Clayton, Costin, Fawcett, Lewis, Park, Plato
convicts, 34, 43, 45, 46, 48, 82, 83
Cook, Captain James, 14
Cooke, H. M. M., 108, 121
Coolringdon, 39, 46, 80, 92, 117, 134
Cooma, 8, 15, 37, 42, 44, 56, 75, 83, 91, 92, 109, 112, 117, 120, 125, 126, 134, 136, 151, 185
Costin, A. B., 170, 171, 174
counties, 8
Crisp, Amos and William, 75, 107–10
Critias, a Platonic dialogue, 58
Crown Lands Alienation Acts, pt. III *passim*
Crown Lands Occupation Acts, pt. III *passim*
  See also free selection; Robertson's Land Acts
Cunningham, Alan, 61
Cuppacumbalong, 128
currency lads, 34

204

Index

Currie, Mark John, 3, 6, 15, 20, 23, 24, 67, 68
Curry Flat, 124

dairying, Monaro ill adapted to, 118, 122
Dangelong, 80, 91, 92, 202
Darwin, Charles, 16, 62
David, Sir Edgeworth, 143, 164
David moraine, 188
Dawson, Robert, 36, 37, 44
Day, M. F., 181
deforestation, 57, 58, 59, 62, 129, 145. *See also* burning, forests, ringbarking
Delegate, 38, 130, 152, 158, 159
Demarr, James, 82, 83
Deniliquin, 91
Descartes, 189
Devereux, James, 96
Donald, C. M., 151
droughts, 37–9, 64, 127, 128, 134, 135
Druitt, Archdeacon, 130
Drysdale, Russell, 14
Dunphy, Myles J., 119

ecology
  the word and the concept, 61–2
  philosophical foundations of, 202
ecological research, 10, 57, 62–3. *See also* Byles, Costin, Day, Fawcett, Hamilton, C.S.I.R.O.
Edgar, Dr Graham, 179
education, 125, 126
Edwards, H. T., 103, 106, 124–6, 135, 155, 156
erosion, 107, 109, 110, 129, 144–7, 169, 170. *See also* Soil Conservation Service
Eucumbene, river, 132, 172

Farrer, William, 119, 127, 128, 152
Fawcett, Stella M., 166
financing land purchase and improvement, 40, 48, 91, 92, 97, 99, 101, 103, 127, 152
fire, *see* burning, bushfires
First Fleet, 10, 12
Flood, Josephine, 131n.
flour mills, 120
forests
  do they bring rain? 58, 63n., 129
  as soil cover, 57–9, 129, 144–6
  multiple use of, 190
  See also Plato, Colbert, Hamilton, Lane Poole, Byles, Stretton
Frankel, Sir Otto, 188
free selection, pt. III 1 *passim*, and maps 9, 10, 11, 12. *See also* Robertson, Morris-Ranken, Litchfield, Pratt, and maps 9, 10, 11, 12, 14
Froggatt, W. W., 111, 113
frontier

'Australia's first', 64n.
  cattle man's, 74, 75
  sheep man's, 73–6
frost hollows, 24

Gare, Neville, 174, 175, 178, 181
Garnock, C. T., 159, 162. *See also* Bukalong, rainfall records
Garnock, George, 81, 159
Geehi, river, 159, 162
Gegedzerick (earlier spelling Gejizrick), 46, 84, 202
Geldmacker, John, 120
Gellimatong (also Jillimatong), 37n.
Geoffrey Hamlyn, by Henry Kingsley, 31–3
'get big or get out', a slogan, 160
Gibson, A., and Gibson's Plains, 132
Ginninderra, 129, 160
  C.S.I.R.O. research station at, 157, 160, 161
Gipps, Governor Sir George, 12, 41, 57, 89
Gippsland, 25, 39, 40, 57, 71, 132, 163
gold, 79, 135
Goodradigbee river, 133
Goulburn, 34, 39, 40, 57, 71, 75, 112
Gould, John, 54
grasslands
  of the world, 23
  of Monaro, 3, 23–7
grazing habits, 64
Gundagai, 132
Gungartan mount, 142
Gungarlin river, 166
Gundrowinga, 15

Hain family, 163
Hain's store, 126
Hale, H., 16, 17
Hamilton, A. G., 62, 63, 110, 111, 114, 117
'hands and knees observation', 172, 174
hares, 177
Harkness, William, 101, 124
Harnett, J., 167
Haslingden family, 163
Haygarth, H. W., 59, 60, 61, 65, 66, 72, 76, 78, 82–5
Hazeldean, 117, 124, 126, 158, 160, 185
Helms, Richard, 19–22, 136, 143, 144, 171, 178
The high country, pt. III 4 and pt. IV 2 *passim*, 187–91
  as used by Aboriginal food gatherers, 21, 22, 131
  and European pasture seekers, 110, 132–6, 139–47, 165, 170–2, 178, 179
  and knowledge seekers, 19–22, 26, 81, 143–6, pt. IV 2 *passim*, 188–9
  and pleasure seekers, 138, 139, 180, 185–8

Index

The high country (cont.)
and power and water seekers, 168–70, 172, 176, 177, 200
and solitude seekers, 174, 178
protection of the catchments, 168–72
and of the native habitat, 178–81, 187–91
Hope, A. D., 12
Hovell, W. H., 59, 73, 152
Howard, Amos, 152
Howitt, A. W., 18n., 19, 25–7, 75
Hudson, Sir William, 169, 170, 177
Humboldt, Alexander von, 38
Hume, Hamilton, 73
Hume-Snowy Bushfire Protection Scheme, 170
Hutchins, Peter, 160, 161

improvement, pts. II 3, III 3, IV 1, passim
concepts of: Macaulay's, 72, 125; Haygarth's, 72, 76. See also acclimatisation
exemplifications of
(1) gardens, 75, 82, 125
(2) houses, 80, 125
(3) fences and station equipment, 80, 122, 123
(4) animal breeding and health, 73, 76, 78, 79, 123, 124
(5) water conservation and supply, 40, 161, 162
(6) pasture, 59, 64, 72, 73, 79, 130, 131, 152, 158–62
(7) soil fertility, 151, 152ff.
(8) economic infrastructure (transport, commercial services, etc.), 75, 79, 80, 92, 126, 128
(9) demographic, 85
(10) educational, 125, 126
(11) moral, 83
incentives and disincentives for
(1) security of tenure, 44, 80, pt. III 1 passim
(2) profitable cost-price relationship, 102, 126, pt. IV, passim, 183–6
(3) family responsibility, 84, 85, 94n.
(4) pride and pleasure in effort and achievement, pts. II 3, III 3 and IV 1 passim
time lag in Monaro, 75, 152, 153
improvements, a term of commerce, 73, 74, 80
'irrigation settlements downstream', 189. See also S.M.A.

Jackson, W. D., 25
Jardine, William, 120, 124
Jauncey, J., 74
Jeffreys family of Delegate, 152, 159

Jemmy Gibber, an Aborigine, 70
Jimenbuen, 107, 112, 117
Jindabyne, 20, 39, 120, 134
de Jouvenal, Bertrand, 183

Kerries, range, 135, 142
Kerry, Charles, 139
Khancoban, 146
Kiandra, 80, 132, 134, 135, 138, 139
Kiandra Snow-Shoe Club, 138, 139
King (George King and Co.), 92
Kingsley, Henry, 31–3, 38
Kiss, W. D., 99
Knockalong, 162
Kosciusko, mount, 20, 55n., 139, 176, 180
Kosciusko State (later National) Park. See Park
Kydra, river, 111
Kydra station, selectors on, 112

labour
bond, 43, 45, 46, 48, 82, 83
free, 48, 70, 77, 82, 83, 124
See also social mobility
Lambie, John, 6, 8, 12, 41–6, 67, 71
Lambrigg, 128
land: title to
tenancy on Crown Land: (a) under local legislation of the 1830s, 42–4; (b) under Imperial Waste Lands Act 1846, 44–53, 89; (c) under legislation of N.S.W. parliament, pt. III, 1 passim. See also snow leases
ownership of: (a) under Imperial Waste Lands Act 1846, 44, 80, 92, 98; (b) under free selection (Robertson) legislation, pt. III 1 passim; (c) under Closer Settlement and Soldier Settlement Acts, 106, 155, 156, 160. See also maps, 7, 9, 10, 11, 12, 14
land, viable areas of, 96, pt. IV 1, passim, 106, 118, 201
land survey, 8, 41n., 89, 90
Lands Department, 99, 136, 139–41, 144, 165, 179
landscapes, the two, 10, pt. IV passim
Lane Poole, C. E., 58n., 144
Lawrence, G. V., 170
Ledger, Charles, 116
Lendenfeld, R. von, 143
Lewin, John, 54
Lewis, Hon. T. L., 168n., 178, 181
Lhotsky, John, 54–6, 66–70, 80–83
Libby, W. F., 17
Limestone Plains, 119, 120
limits of location, 8, 41, 42. See also Nineteen Counties; survey.

206

Index

Index

Index